T0335553

DOING SCIENCE

In the Light of Philosophy

DOING SCIENCE
In the Light of Philosophy

Mario Bunge
McGill University, Canada

A.Ibáñez et al. on *Free Will*
M. Mahner on *Philosophy of Mind*

 **World Scientific**

NEW JERSEY · LONDON · SINGAPORE · BEIJING · SHANGHAI · HONG KONG · TAIPEI · CHENNAI · TOKYO

Published by

World Scientific Publishing Co. Pte. Ltd.
5 Toh Tuck Link, Singapore 596224
USA office: 27 Warren Street, Suite 401-402, Hackensack, NJ 07601
UK office: 57 Shelton Street, Covent Garden, London WC2H 9HE

Library of Congress Cataloging-in-Publication Data
Names: Bunge, Mario, 1919–
Title: Doing science : in the light of philosophy / by Mario Augusto Bunge
    (McGill University, Canada).
Description: New Jersey : World Scientific, 2016. | Includes bibliographical
    references and indexes.
Identifiers: LCCN 2016040513 | ISBN 9789813202764 (hardcover : alk. paper)
Subjects: LCSH: Science--Philosophy. | Science--Methodology.
Classification: LCC Q175 .B82228 2016 | DDC 501--dc23
LC record available at https://lccn.loc.gov/2016040513

British Library Cataloguing-in-Publication Data
A catalogue record for this book is available from the British Library.

Printed in Singapore

PREFACE

The contemporary science studies investigate most of the facets of science, but they tend to focus on answers at the expense of questions. They also neglect the influence of philosophy on the problematics, methodics, and evaluation of scientific research. The present investigation seeks to overcome both limitations, by focusing on the research project nurtured or thwarted by the philosophical matrix that hosts it.

Another goal of the present investigation is to restore the classical view of scientific research as the search for original truths. This view was seriously challenged in the 1960s by the opinions of Thomas Kuhn and Paul Feyerabend, that scientists do not seek truth because there is no such thing; of Bruno Latour and his fellow constructivist-relativists, that scientists make up facts instead of studying them objectively; and of Michel Foucault, that "science is politics by other means."

However, I am not out to revive the traditional view of science as a repository of reliable data, and even less to defend Popper's extravagant opinion that scientists are masochists bent on falsifying their own pet hypotheses. Although this book abounds in criticisms of the most popular opinions of science, its central thrust is to proceed with the constructive task of proposing the new theory of scientific research that I started in my two-volume work *Scientific Research* (1967b).

The present theory includes and refines, among others, the concept of indicator or marker, absent from the empiricist accounts of measurement; it also includes and refines the concept of reference, absent from the best-known semantical theories, such as Carnap's, which confuse meaning with testability. In addition, the new theory avoids the confusion of measurement — an empirical operation — with the set-theoretic concept of measure; it also avoids the confusion of the dimension of a magnitude, such as LT^{-1} in the case of velocity, with that of size; last, but not least, the present view, contrary to the structuralist one, adopts the cutlet model of a scientific theory as a mathematical formalism endowed with a factual content or set of semantical hypotheses.

In short, the main thrust of the present work is to propose a view of scientific research as it is actually conducted by active scientists. Interestingly, this effort to get closer to science in the making also gets us closer to philosophy than the standard views. I will argue that this philosophical matrix of scientific research performs both heuristic and regulative roles, and that it constitutes a whole worldview that is hoped to fit in with contemporary science.

To the extent that it satisfies this realism condition, that rather tacit worldview deserves being called *scientific*. Far from being an intellectual game, this particular mode of looking at science is expected to help us flag down the beliefs and practices that, like the alternative medicines and the sectoral and opportunistic social policies, contradict the so-called spirit of science and ignore the relevant scientific evidence, whence they constitute public perils. Up to a point, this trait vindicates the ancient view of philosophy as a guide to life, as well as Aristotle's view of science as a single body of perfectible knowledge.

The main body of this book is followed by two appendices about the science and philosophy of mind. Appendix 1, by Professor Facundo Manes and some of his coworkers in neuroscience, concerns one of the Big Questions, namely the existence of free will. This question, first explored in 400 C.E. by St. Augustine of Hippo, has been the subject of heated but inconclusive debates ever since. Most scientists, supporters of a narrow version of determinism, have rejected free will as one more theological fantasy. So have the defenders of the computer metaphor of the mind. By contrast, Donald Hebb (1980), the founder of theoretical neuropsychology, was also the first modern scientist to suggest that it is a legitimate subject of experimental psychology. This is also the way the authors of Appendix 1 view it, and they examine a pile of fascinating recent neuroscientific findings relevant to the subject. (See also Burge 1980.)

The author of Appendix 2, the zoologist Dr. Martin Mahner, concurs with the scientific approach to the mental, and does his best to clarify some of the key philosophico-scientific terms occurring in present-day philosophy of mind. He also shows that the confusions that plague this important chapter of metaphysics or ontology have hindered the advancement of the sciences of mind. Mahner also suggests that the prevailing philosophies of mind lag behind the corresponding science, whereas my own contributions to the field have accompanied and helped psychological research.

Mario Bunge
Department of Philosophy
McGill University
Montréal, Canada

CONTENTS

INTRODUCTION

This book focuses on science in the making as well as on its philosophical presuppositions, such as those of rationality and realism. Although these presuppositions are mostly tacit and thus easily overlooked, actually they are supremely important, since some of them favor research whereas others hamper it. For instance, whereas subjectivism leads to navel gazing and uncontrolled fantasy, realism encourages us to explore the world and check our conjectures.

The bits of science we learn in schools and textbooks are finished products, whereas the results of recent scientific projects are published in journals accessible only to specialists. Thus, the American Institute of Physics alone publishes 19 peer-reviewed journals. On the occasion of his visit to that institute, the famous professor of philosophy of science at Princeton University was amazed to learn that there was more than one physics journal in the world. Obviously, he did not consult science journals.

Scientific journals publish original reports, review articles, and short notices. The rank of scientists is roughly gauged by the number of papers published in high-impact journals — a debatable metric for, as John Garcia's (1981) sensational experiment showed, it conflates research quality with the prestige of the author's academic home. Some of those journals have such high standards, that they publish just one in 10 or more submissions. The popular press spreads only rumors about a few outstanding papers.

Evidently, reading recent papers in prestigious science journals is not enough to train productive investigators. Science in the making is learned only by doing or replicating some original scientific research, and even so provided one succeeds in piercing the thick layers of myths about science, such as its confusion with technology or even with the search for power (see Numbers & Kampourakis 2015 for a representative sample).

Take, for instance, the difference between scientific thought and free fantasy. While it is true that uncontrolled fantasy belongs in literature, art, and theology, it is also true that scientifically rigorous work without fantasy is just one more routine job, like cooking, sewing, or computing in accordance with given rules. Fantasy — the flight from the obvious or the wellworn — is of the essence of original work, whether in science, technology, art, literature, management, or daily life. However, let us not be carried away by the similarities, because scientists search for truth, which is optional in other fields.

That, the centrality of fantasy in science, is why the 19th-century German university bureaucracy used to pack mathematicians and theologians into a single division. Of course, by so doing they overlooked the point that, unlike theologians, mathematicians spend most of their time proving conjectures rather than making them up. But proving comes only after conjecturing, and there are no rules for coming up with new hypotheses: there is no such thing as an *ars inveniendi*.

Another popular mistake concerns the impersonality of science. When stating that, unlike love and taste, scientific research is impersonal, one intends to say that scientific procedures and outcomes are scrutable and subject to appraisals in accordance with objective criteria, such as originality, clarity, precision, consistency, objectivity, compatibility with the bulk of antecedent knowledge, replicability, and belonging in the public domain rather than privy only to a sect.

This has not always been the case. At the time when universities started to be divided into departments rather than chairs, the latter's holders kept behaving like feudal lords, and some of them claimed to own their research areas. In the 1960s, when the West European universities were regrouped into departments, a prominent Heidelberg professor, to be called M-L, refused to comply, and placed on his door a plate that read *M-L Abteilung*. And in a Spanish university, the specialist in dimensional analysis — a tiny and exhausted chapter of applied mathematics — got himself appointed as the one-man *Departamento de Análisis Dimensional*.

As for the science-philosophy connection, take, for instance, the study of *qualia* or secondary properties, such as color, taste, and smell. When I was attending high school in the 1930s, the study of chemistry included memorizing the organoleptic properties of the various substances we studied in the textbook but never handled. For example, we had to learn that chlorine looks yellowish-green, tastes mordant, and has a choking smell, all of which was true and useful in the lab, but did not contribute to understanding chemistry any more than learning how to remove stains from lab coats.

Most philosophers have assured us that, since all of these properties (the *qualia*) are subjective, none of them could possibly be accounted for by science. Indeed, as Galileo had taught four centuries earlier in *Il saggiatore* (1693), science (then meaning mechanics) deals only with primary properties such as shape, heaviness, and speed. Even today, the very existence of qualia is often used to refute materialism, vulgarly identified with physicalism.

Those philosophers would be astonished to learn that qualia are nowadays being analyzed into primary properties of certain brain subsystems (see, e.g., Peng *et al.* 2015). This study is being conducted by cognitive neuroscientists who, using brain-imaging techniques, have come up with a whole gustotopic map of taste qualities in the "mammalian" (actually just murine) brain. In particular, they have located taste in the insula — an organ located under the brain cortex and near the eyes. They have learned that the sweet and the bitter percepts are separated by approximately 2 mm. Such qualia can be excited not only by food ingestion but also either by photostimulation or by injecting certain drugs. Moreover, not having read nativists like Noam Chomsky or Steven Pinker, mice can be trained to overcome the innate drive, which is preference for sweet over bitter cues.

All of these investigations have presupposed the materialist (though non-physicalist) hypothesis that everything mental is cerebral. None of them would have been even contemplated if the investigators had remained shackled to spiritualism, psychoneural dualism, or computerism — the three philosophies of mind currently favored by most philosophers of mind.

Lastly, the current scientific investigation of qualia is not only an example of the *Philosophy* → *Science* action. It also shows the occurrence of the converse action: that some scientific findings can force certain philosophical changes — in this case the enlargement of materialism to encompass subjective experience and its objective study, even that of free will (see Appendix 1).

More precisely, ancient materialism, born as mechanism in both Greece and India two and a half millennia ago, is now just one smallish sector of scientific materialism, which can influence all of the scientific disciplines as long as it tames the mental instead of denying it. For example, the popular claim that intention differs radically from causation evaporates when learning that intentions are processes in the prefrontal cortex.

Nearly all of the above falls under the philosophy of science. This discipline is not much respected by most scientists. Thus Richard Feynman once quipped that the philosophy of physics is as useful to physicists as ornithology is to birds. One may reply that scientists cannot help philosophize, as when they wonder about the real occurrence and scrutability of some hypothesized entity or process. For example, the whole point of building and financing huge particle accelerators like those in CERN, Fermilab, and Dubna, is to find out whether some of the entities and events imagined by theorists are real (see Galison 1987).

Moreover, had Feynman paid some attention to philosophy, he would not have confused laws (objective patterns) with rules (prescriptions for doing things); he would not have assumed that positrons are electrons moving to the past; he would have regarded his own famous graphs as mnemonic devices rather than as depictions of real trajectories; and he would not have written that, "Since we can write down the solution to any physical problem, we have a complete theory which could stand by itself" (Feynman 1949).

Feynman could afford pushing philosophy aside, ignore the great Bohr–Einstein debate on realism, and declare that "no one understands quantum mechanics," for he had chosen to perform a lot of very hard calculations — a task that demands no philosophical commitment. Besides, Feynman was working in a mature branch of physics, namely electrodynamics, fathered a century earlier by André-Marie Ampère, who in 1843 had published a two-volume work on the philosophy of science.

By contrast, Charles Darwin knew that he had started a new science, which had to be protected from the attacks of the conservative establishment. This is why he wore the ruling philosophical mask in public, while confiding his heterodox beliefs only to a few close friends and his private *Notebooks M and N* (Ayala 2016; Gruber & Barrett 1974). There we learn of

Darwin's religious skepticism, his materialist philosophy of mind (in 1838!), and his non-empiricist theory of knowledge. In particular, Darwin held that — contrary to what the British empiricists taught — every useful (nontrivial) scientific observation is guided by some hypothesis. Such philosophical heterodoxies are likely to have helped Darwin devise his scientific ones and, above all, carry out his grand goal of revealing the tree of life.

CHAPTER 1

IN THE BEGINNING WAS THE PROBLEM

To engage in research of any kind is to work on a problem or a cluster of problems of some kind — cognitive, technological, social, artistic, or moral. In imitation of John's gospel, we may say that *in the beginning was the problem*. So, those wishing to start doing science must find or invent a problem to work on, as well as a mentor willing to guide them.

1.1 At the Source

Free agents prefer to work on problems they like and feel are equipped to tackle. But of course most budding scientists are not fully free to choose: their supervisors or employers will assign them their tasks — for one does not know one's own ability before trying and, above all, because finding a suitable problem is the first and hardest step.

However, problem choice is only part of a whole package, which includes also such noncognitive items as advisor's suitability and availability, research facilities, and financial assistance. In other words, the budding scientist or technologist does not enjoy the luxury of picking his/her favorite problem — which is just as well because, given his/her inexperience, that choice is likely to be either too ambitious or too humble. In sum, aspiring investigators are given to choose among a set of packages offered by his/her prospective advisor or employer.

For better or for worse, there are no recipes or algorithms for generating problems other than reviewing the recent literature. In particular, computers cannot pose problems, for they are designed, built, and sold to help solve well-posed problems, such as curve fitting a given set of data points. After listening to Stanislav Ulam's panegyric of the abilities of computers, I left him speechless by asking him, at a congress packed with sages, whether such marvels might invent new problems. He paused for a long while and finally admitted that this question had never occurred to him. Such is the power of raw data and data-processing devices.

Half a century ago, Alan Turing proposed the test that bears his name as the way to discover whether one's interlocutor is a human or a robot. Later work in AI showed that Turing's test is not foolproof. There is an alternative: ask your interlocutor to pose a new and interesting question. Computers will fail this test, because they are designed to operate on algorithms, not to deal with questions that demand invention, in particular problems, such as guessing intention from behavior. This test is therefore one about natural intelligence, or thinking out of the digital box.

1.2 Types of Problems

The logical positivists like Philipp Frank, as well as their critic Karl Popper, banned questions of the "What-is-it?" type. By contrast, the great physiologist Ivan Pavlov (1927:12) held that they exemplify what he called the *investigatory reflex*, for they elicit an animal's response to environmental changes. Indeed, they constitute existential dilemmas, hence the most basic of all, for they include "Friend or foe?," "Safe or risky?," "Edible or inedible?," and the like.

Admittedly, only humans and apes capable of communicating with us via sign language or computers will formulate problems in a sentence-like fashion. But this is a moot point: what matters mostly is that the animals that do not solve their existential dilemmas are unlikely to survive — unless they are tenured philosophy professors.

The importance of problems in all walks of life is such, that someone said that living is basically tackling problems. For those who have solved the subsistence problem, to live is to fall in and out of love with cognitive, valuational, or moral problems. Those of us who ask Big Questions, such as "How and why did civilization start?" are called bold scientists. And the few who ask the biggest questions of all, such as "What exists by itself?," "What is truth?," and "Is science morally neutral?" are called philosophers. These and similar questions are transdisciplinary, whereas all the others are unidisciplinary.

1.3 Erotetics

Most philosophers have overlooked problems and their logic, namely erotetics, which should be the subject of countless original philosophical research projects. The next few pages will recall the author's erotetics discussed in what is likely to have been the first treatise in the philosophy of science to sketch it (Bunge 1967b, vol. 1).

Whatever the kind of cognitive problem, we may distinguish the following aspects of it: (a) the statement of the problem regarded as a member of a particular epistemological category; (b) the act of questioning — a psychological subject; and (c) the expression of the problem by a set of interrogatives or imperatives (the linguistic aspect). In the present section we shall focus on the first of these aspects.

From an action-theoretic viewpoint, a problem is the first link of a chain: *Problem — Search — Solution — Check.* From a logical point of view, the first link may be analyzed into the following quadruple: *background, generator, solution* (in case it exists), and *control* — or *BGSC* for short.

Let us clarify the preceding by way of a "hot" example in astrophysics, namely "What is dark matter?" We start by reformulating the given problem as "Which are the properties P of the Ds?," or $(?P)Px$, where x designates an arbitrary member of the class D of all possible pieces of dark matter, and P a conjunction of known and new physical properties. The BGSC components of this particular problem are

Background B = Contemporary astrophysics plus particle physics.
Generator G = Px, where P = a conjunction of first-order properties.
Solution S = The cluster P of properties assignable to any D.
Control C = The laboratory analysis of a piece of dark matter or of the radiation (other than light) that it emits.

Let us close by listing the elementary problem forms.

Which-problems Which is (are) the x such that Px?	$(?x)Px$
What-problems Which are the properties of item c?	$(?P)Pc$
How-problems How does c, which is an A, happen?	$(?P)[Ac \Rightarrow Pc]$
Why-problems Which is the p such that q?	$(?p)(p \Rightarrow q)$
Whether-problems What is the truth-value of p?	$(?v)[V(p) = v]$
Inverse problems Given B and $A \rightarrow B$, find A.	$(A?)[A \rightarrow B]$

The direct/inverse distinction may be summarized thus:

| Direct | Input → System → Output |
| Inverse | Output ← System ← Input |

In the simplest case, the input–output relation is functional, and it can be depicted as follows:

| Direct | $x \to f \to f(x)$ |
| Inverse | $f(x) \to f^{-1} \to x$ |

However, most real-life problems of are of the means-end kind, most of which have multiple solutions, so they are not functional.

Whereas direct problems are downstream, or from either causes or premises to effects or conclusions, the inverse ones are upstream, or from effects or theorems to causes or premises. A common inverse problem is that of conjecturing a probability distribution from statistics such as average and mean standard deviation. A far less common inverse problem is the axiomatization of a theory known in its ordinary untidy version (see Chapter 7).

Like most inverse problems, axiomatics has multiple solutions. The choice among them is largely a matter of convenience, taste, or philosophy. For example, whereas an empiricist is likely to start with electric current densities and field intensities, the rationalist is likely to prefer starting with current densities and electromagnetic potentials, because the latter imply the field intensities (see Bunge 2014; Hilbert 1918).

All the prognosis problems, whether in medicine or elsewhere, are direct, whereas the diagnostic ones are inverse. For example, having diagnosed a patient as suffering from a given disease on the strength of a few symptoms, checking for the occurrence of further symptoms is a direct problem. But the problem of medical diagnosis is inverse, hence far harder, for it consists in guessing the disease from some of its symptoms.

Ordinary logic and computer algorithms have been designed to handle direct problems. Inverse problems require inventing ad-hoc tricks, and such problems have multiple solutions or none. For example, whereas 2 + 3 = 5, the corresponding inverse problem of analyzing 5 into the sum of two integers has four solutions.

Inverse problems may be restated thus: given the output of a system, find its input, mechanism of action, or both. That is, knowing or guessing that $A \rightarrow B$, as well as the output B of a system, find its input A or the mechanism M that converts A into B. For example, given a proposition, find the premises that entail it; design an artifact that will produce a desired effect; and given the beam of particles scattered by an atomic nucleus, guess the latter's composition, as well as the nature of the scattering force (For the pitfalls of this task see, e.g., Bunge 1973a).

A fever may be due to umpteen causes, and its cure may be achieved through multiples therapies, which is why both biomedical research and medical practice are so hard (see Bunge 2013). As a matter of fact, most inverse problems are hard because there are no algorithms for tackling them. This is why most philosophers have never heard of them. The referees of my first philosophical paper on the subject rejected it even while admitting that they had never encountered the expression 'inverse problem' (Bunge 2006).

Finally let us ask whether there are insoluble problems. Around 1900 David Hilbert stated his conviction that all well-posed mathematical problems are soluble in principle, not just unsolved up to now. Here we shall disregard unsolvable mathematical problems because they are arcane questions in the foundations of mathematics, and anyway they have raised no philosophical eyebrows. We shall confine ourselves to noting that some seemingly profound philosophical problems are ill posed because they presuppose a questionable background.

The oldest and most famous of them is, "Why is there something rather than nothing?" Obviously, this question makes sense only in a theodicy that supposes that the Deity, being omnipotent, had the power of inaction before setting out to build the universe: why bother with real existents if He could spend all eternity in leisure? Taken out of its original theological context, the said question is seen to be a pseudoproblem, hence not one that will kindle a scientific research project. In a secular context we take the existence of the world for granted, and ask only particular existence problems, such as "Why do humans have nails on their toes?," which is asked and answered by evolutionary biologists. The answer is of course that toenails descend from the fingernails that our remote ancestors had on their hind legs, which worked as hands.

Yet it is often forgotten that all problems are posed against some context, and that they vanish if the context is shown to be wrong. Let us recall a couple of famous games of this kind.

Pseudoproblem 1: What would happen if suddenly all the distances in the universe were halved? The answer is in two parts: (a) nothing at all would happen, for all distances are relative, in particular relative to some length standard, which would also shrink along with everything else; and (b) since no universal shrinkage mechanism is known in physics, the said event should be regarded as miraculous, hence conceivable but physically impossible.

Pseudoproblem 2: What is the probability that the next bird we spot is a falcon, or that the next person we meet is the pope? Answer: neither belonging to a given biospecies nor holding a particular office are random events, so the given questions should be completed by adding the clause "picked at random from a given population (of birds or people respectively)." No randomness, no applied probability. In conclusion, unless the background of a question is mentioned explicitly, it won't start a research project.

A final warning: genuine cognitive problems are not word games played just to exercise or display wit. The best known of these games is perhaps the Liar Paradox, generated by the sentence 'This sentence is false.' If the sentence is true, then it is false; but if it is false, then it is true.

The paradox dissolves either on noticing that the sentence in question conflates language with metalanguage; or that it does not express a proposition, for propositions have definite truth-values.

The first interpretation warns against such confusions, and the second reminds us that only propositions can be assigned truth values, whence it is wrong to call first-order logic 'sentential calculus', the way nominalists do just because of their suspicion of unobservables. In sum, avoid barren paradoxes when stating a cognitive problem, for truth is not a toy.

1.4 The Search for Research Problems

How does one find a suitable research problem? The answer depends on the kind of problem: is it a matter of survival, like finding the next meal ticket; a technological problem, such as increasing the efficiency of an engine; a moral problem, such as how to help someone; or an epistemic problem, such as to discover how dark holes arise or evolve?

The question of problem choice has mobilized psychologists, historians, and sociologists. These experts have attacked what Thomas Kuhn (1977) called 'the essential tension.' This is the choice between a potboiler that may inflate the investigator's CV but won't alter anyone else's sleep, and a risky adventure with an uncertain outcome that may alter an important component of the prevailing worldview, as was the case when Michael Faraday assumed that electric charges and currents, as well as magnets, interact via massless fields rather than directly.

Familiar examples of the first kind are spotting a previously unknown celestial body of a known kind, the chemical analysis of a newly discovered wild plant, and computing or measuring a well-known parameter with greater precision. In contrast, looking for evidence of the ninth planet of our solar system, digging for hominid fossils in a newly found archaeological site, and searching for a better cancer therapy, are instances of long-term and risky projects. They are risky in the sense that one embarks on them even while fearing of wasting time and resources.

Unsurprisingly, in every walk of life traditionalists outnumber innovators. However, though real, the conservative/innovative "tension" is transient, since the initial success of a groundbreaking research project is bound to attract droves of researchers who inaugurate a new tradition.

Occasionally, even the news that an established scientist is trying a new approach would have the same result, namely the sudden recruitment of hundreds of young researchers working on the same project. This used to happen in particle physics in the 1950s and 1960s, when certain new theories became instantly fashionable for a few months. Some of the most ambitious projects attracted exceptionally able investigators, and remained in fashion even if they failed to deliver the promised goods. String theory is one of them. Nowadays this theory resembles the aging hippie who keeps wearing his jaded jeans.

However, the conservative/innovative distinction is best drawn once the problems concerned have been worked out. The first question one should tackle is about the main sources of problems, and this question is tacitly answered the moment a problem typology is proposed. For example, moral, political, and legal problems arise only in the course of social interaction. Thus, Robinson Crusoe felt no moral qualms before meeting Friday. As soon as this meeting occurred, each of them must have asked himself how best to treat the other: as friend, foe, or neither; as competitor, cooperator, or neither — and so on.

In contrast, pure curiosity prompts us to asking epistemic questions, such as whether the squared root of 2 can be expressed as the ratio of two whole numbers; whether dark matter is anything other than clumps of matter whose constituent atoms have fallen to their lowest energy level; and whether primitive living cells might soon be synthesized in the laboratory.

Collecting problems into a number of different boxes or kinds sparks off yet another problem, namely that of the possible links among such boxes. A familiar member of this kind is the relation between science and technology. The standard answer to this question is that science generates technology, which in turn poses scientific problems, so that each feeds the other. Let us briefly consider the four most popular answers, and then a fifth, namely whether philosophy too can meddle with science, now helping, now obstructing it.

So far, the corresponding findings of the historians and sociologists of science, technology, and philosophy on the above questions constitute a motley collection of isolated items that are so many problems. Suffice it to list the following famous items:

a. What led to the discovery of "irrational" numbers, that is, numbers that are not ratios of integers like 2/3? Short answer: the wish to corroborate Pythagoras's postulate, that the basic constituents of the universe are whole numbers. According to legend, the member of his fraternity who dared disputing this conjecture, to the point of proving that the square root of 2 is irrational, was put to death. In short, his research project had a philosophical motivation.

b. Why did some ancient Greek and Indian thinkers hold that all things in the world are combinations of bits of fundamental or indivisible things? Perhaps because carpenters, bricklayers, and other craftsmen make artifacts by assembling or dividing things. Those were deep thinkers, not problem-solvers, interested only in ordinary life issues. Parallel: the metamathematician who asks not what the result of a computation is, but what kind of animal a computation is.

c. Why did Erathostenes wonder about the shape and size of planet Earth? Presumably, because he was curious and examined some of the indicators that our planet was not flat, such as the seeming sinking of westbound ships.

d. What led Empedocles to suspect that the biospecies had evolved instead of being fixed? Perhaps he was led by religious skepticism jointly with the finding of marine fossils on the top of certain mountains.

e. What led Olaf Rømer to design his ingenious device to measure the speed of light in 1676, when light was generally regarded to travel at infinite speed? Likely, it was his observation that Jupiter's eclipses were seen at different times at different times of the year, when that planet was at two different places. So, light had to take some time to travel from Jupiter to Earth. His was a case where a research project sprang from a surprising observation. Likewise, in 1820 Johannes Müller undertook to measure the speed of nervous pulses out of sheer curiosity — the main fountain of science according to Aristotle.

f. Why did Newton try out the inverse square hypothesis? As told by the diarist Samuel Pepys, a number of amateurs who used to meet at a coffee house surmised that our solar system was held together by an unknown force, and one of the habitués offered a substantial prize to the first man to propose a plausible solution. Newton had just crafted the first theory that allowed the exact statement of this inverse problem, which he solved by transforming it into a direct problem: calculate the planet trajectory assuming that the force binding it to the sun is inversely proportional to the square of their mutual distance. This, Newton's great inverse problem, rather than Hume's problem of induction, is the one that started a new epoch in theoretical science.

g. In 1688 the physician Francesco Redi put to the test the popular hypothesis of spontaneous generation, by isolating a morsel of meat. Contrary to expectations, no flies appeared; they only came from larvae that deposited on rotting meat. Lazzaro Spallanzani and others corroborated this result during the next three centuries: over that period the maxim *omne vivo ex vivo* held sway. The materialist hypothesis (abiogenesis) that the earliest organisms had been born spontaneously (without God's help) from the synthesis of abiotic materials was regarded as having been falsified once and for all.

h. In 1862 Louis Pasteur, the founder of microbiology, endeavored to find out whether his microbes might develop into more complex organisms. To this end he made a jar of sterile nutrient broth, which he boiled. That is, he unwittingly made sure that no life germs remained. Unsurprisingly, microscopic observation showed no signs of life. Once again, experiment was regarded as having killed the myth of spontaneous generation. At the time no one dared analyze Pasteur's experimental design. He succeeded in falsifying a myth — for a while.

i. In 1953 the chemistry graduate student Stanley Miller and his mentor, the seasoned physicist Harold Urey, synthesized some amino acids and other organic molecules out of a mixture of methane, ammonia, water, and hydrogen, which they subjected to electric discharges on the assumption that the primitive atmosphere had been similarly electrified. True, the outcome was inconclusive, for no living things were produced. But suddenly the synthesis of life became a serious research project, and it did so propelled by the philosophical hypothesis that life might indeed have emerged from lifeless precursors — as any good materialist would surmise.

After more than one century, the Miller–Urey project is still been pursued even though their result had fallen into the crack between the two traditional categories, *confirmed* and *falsified*, as shown in the following diagram.

Confirmed
Experimental outcome / Inconclusive
\ Falsified

We shall return to this problem in Chapter 3. Suffice it for now to note that the standard philosophies of science, in particular confirmationism and falsificatonism, take it for granted that (a) all experimental results are unequivocal, and (b) experiment is the undisputed umpire that arbitrates between competing hypotheses. Clearly, these two pillars of the standard philosophy of science are cracked. And assuming that hypotheses can only be more or less "probable" won't repair those pillars, for assigning hypotheses subjective (or personal) probabilities is unscientific (Bunge 2008).

1.5 Problem System

It is well known that problems come in all kinds and sizes. There are cognitive and moral problems, individual and social issues, scientific and technological conundrums, political issues, and so. Then again, problems can be either local or systemic, and they may be tackled either by individual experts or by multidisciplinary teams.

Small problems call for the use of known tools found in circumscribed fields, whereas big problems call for further research, which may require breaching disciplinary walls. For example, whereas an experienced bone-setter may fix a fractured bone, a problem concerning invisible entities may require interdisciplinary research.

Another philosophically significant partition of the set of scientific problems is the direct/inverse dichotomy. The problem of induction is the best known of all the inverse philosophical problems. It consists in leaping from a bunch of data to a universal generalization. For example, given a bunch of scattered dots spread on a Cartesian grid, find a smooth curve joining them. As every calculus student knows, the standard solution to this problem is the interpolation formula invented by James Gregory in 1670. This is an nth power polynomial $f(x)$ constructed from a set of $n + 1$ values of f plus the smoothness assumption.

Gregory's formula is only good to handle low-level variables, such as the stretching and load of spring scales (Hooke's law), and Ohm's law, which relates the voltage and current intensity of a direct-current electrical circuit. The higher-level law statements, such as those occurring in electrodynamics, cannot be reached by induction because they go far beyond data.

The inception of Gregory's curve-fitting method was neither inductive nor deductive: it was an invention — or abduction, as Charles Peirce might have called it. Only the problem that prompted it, namely going from a bunch of data to a general formula, deserves being put under the heading of induction.

Incidentally, nowadays Gregory's invention is usually called the Newton–Gregory formula. This is an instance of the Matthew effect, studied by Robert Merton, whereby a minor scientist is hitched to a famous one in order to highlight the finding's importance. Another, far less important case, is that of the Feynman–Bunge coordinate.

Going back to Hume, he is unlikely to have heard of the problem in question, much less of Gregory's solution to it. Nor did Popper, who three centuries later claimed to have solved what he called 'Hume's problem.' Note that Gregory's formula is only good to handle low-level variables, such as the stretching and load of spring scales (Hooke's law). Higher-level law statements, such as those occurring in electrodynamics, cannot be reached by induction because they go far beyond data.

The hardest problems are the so-called Big Questions, such as the one concerning the origin of life. As reported above, in 1953 Harold Urey and Stanley Miller approached this problem from scratch, that is, from simple molecules such as hydrogen, water, methane, and ammonia, they did not get living things but instead nucleotides–essential components of the DNA molecule. Thus, Miller and Urey had produced the first solid result in synthetic biology, after Aleksander Oparin's bold and inspiring work in the 1920s. The same year 1953, help came from an unexpected quarter: the molecular biology due to Francis Crick and James Watson. This breakthrough suggested an alternative strategy for creating life in the lab, namely starting from highly complex organic molecules, such as nucleosides, instead of proceeding from scratch.

This new strategy soon yielded some sensational results, such as Har Gobind Khorana's synthesis of a gene in 1972, and Craig Venter's synthesis of the whole genome of a bacterium in 2010. Although synthetic life is still only a promise, no biologist doubts that it will be achieved in the foreseeable future by following the method of building increasingly complex systems by joining lower-level entities.

To conclude, note that the standard philosophies of science, in particular confirmationism and falsificatonism, take it for granted that (a) all experimental results are unequivocal, and (b) experiment is the undisputed umpire that arbitrates between competing hypotheses. Clearly, these two pillars of the standard philosophies of science are cracked: some experimental results are inconclusive, and a research program may be pursued despite setbacks, as long as it is backed by a strong philosophical hypothesis, such as that of abiogenesis.

CHAPTER 2

SCIENTIFIC RESEARCH PROJECTS

Original research is of course what scientists are expected to do. Therefore the research project is in many ways the unit of science in the making. It is also the means towards the culmination of the scientists' specific activities and the reward for their work: the original publication they hope to contribute to the scientific literature. The scientific project should therefore be of central interest to all the students of science, particularly the philosophers, historians, and sociologists of science.

In the following we shall focus on the preliminary evaluation of research projects — the specific task of referees — and will emphasize the problem of their scientific status — the chief concern of scientific gate-keepers. In the past such an examination aimed only at protecting the taxpayer from swindlers and incompetent amateurs, such as the inventors of continuous motion machines.

In recent times a similar issue has resurfaced with regard to some of the most prestigious and most handsomely funded projects, namely fantasies about string theory and parallel worlds. Indeed, some of their faithful have claimed that these theories are so elegant, and so full of high-grade mathematics, that they should be exempted from empirical tests. As Dirac once suggested, "pretty mathematics" should suffice.

This claim provoked the spirited rebuttal of the well-known cosmologists George Ellis and Joseph Silk (2014), which the present book is intended to reinforce. Indeed, we shall try to show why empirical testability is necessary though insufficient for a piece of work to qualify as scientific. Contrary to the popular theses, that either confirmability or falsifiabiliy is the mark of science, I submit that the most reliable scientificity indicator is the combination of precision with testability and compatibility with the bulk of antecedent scientific knowledge. Indeed, ordinarily science gatekeepers, such as journal and funding agency referees, do not waste their time with imprecise guesses, untestable and at variance with all that is well known.

Finally, the present chapter may also be regarded as an indirect contribution to the current debate over the reliability of quantitative indicators of scientific worth, such as the *h*-index of scientific productivity (e.g., Wilsdon 2015). However, we shall touch only tangentially on the sociology, politics, and economics of research teams: our focus will be the acquisition and assessment of new scientific knowledge through research.

2.1 Scientists Work on Research Projects

All active scientists work on at least one research project at a time. The ultimate goal of any such project is to find new truths or new procedures in the cases of basic and applied science, and new useful artifacts in that of technology. When the finding is interesting but very different from the original goal, one calls it a case of lucky finding or serendipity — like Friedrich Wöhler's when in 1828 he synthesized urea, then believed to be present only in organisms, while trying to obtain ammonium cyanate from carbon monoxide and the amino group. (For serendipity see Merton & Barber 2004.)

At any given time, investigators work hands-on on their pet project, as well as hands-off on several others via their doctoral and postdoctoral students. Normally, every such project is funded by government agencies or private firms in response to the research proposals submitted by the principal investigator. Research grants are awarded to projects to be carried out by individuals, teams, or institutes, provided they pass the scrutiny of a panel of experts. The keys to this test are the originality, soundness, and feasibility of the project together with the ability of its proponents. No panel of rigorous judges would have passed Richard Dawkins's selfish-gene fantasy, Noam Chomsky's revival of the innate knowledge dogma, or Steven Pinker's thesis (2002) that social standing is in the genome, for none of these opinions have resulted from any research projects in either developmental biology or experimental psychology. The same holds for the wild speculations of information-processing psychologists, sociobiologists, pop evolutionary psychologists, many-worlds cosmologists, and alternative medicine practitioners.

There is no live science without ongoing well-grounded scientific projects. This is why ill-founded research projects are usually failed upon submission — even though there is no consensus on this matter, which will be discussed below. The same consideration is behind the advice of experienced researchers, to pay as much attention to the writing of a grant proposal as to that of the final report on its findings.

It has been said, with tongue in cheek, that the four to six weeks it takes to write a grant proposal is the time of year that researchers think really hard: once the problem has been formulated, and the means to solve it have been identified and assembled, the rest is a matter of more or less sophisticated routine. Paraphrasing what Buffon famously said of genius, the conception of a project takes inspiration, while its execution takes only perspiration. But, of course, investigation may show that the original project has to be redone, reconceived from scratch, or even aborted.

2.2 Research Team

So far, we have dealt only with the intrinsic worth of scientific research projects. The institutional aspect is very different, and it will be treated only tangentially: a poor but well-written project may get support, whereas an important but ill-presented one may be turned down by a funding agency. Regrettably, connections and politics still play a role in the evaluation and execution of research proposals.

To be on the safe side, the granting agency should look not just at the project but also at the proponent's track record, as well as at the standing of the experts asked to review it, together with the worth of competing projects. If a project has no competitors, it is likely to be either very original or a piece of quackery, whereas if it has plenty of competitors, it is likely to be either a potboiler or a fashionable fantasy.

Most scientific papers produced by research teams claim that "all the team members contributed equally." Only seldom is the full truth revealed: that the team leader A conceived the project, his collaborator B planned it, C initiated it, coworker D developed the key methods, E assembled or prepared the materials, F collected the data, G processed them, H helped with the operations, the lab manager J took care of the nitty-gritty, K offered valuable advice, and L wrote the paper "with contributions from all authors."

In other words, *the research team as a whole executes the project conceived by its principal investigator(s)*. The technical benefits of the division of scientific labor are as obvious as its psychological and social pitfalls — usually only the leader and his/her close associates have a full grasp of the project as a whole, and the assistants may feel marginal to it because they are replaceable.

Teamwork and *esprit de corps* are of course indispensable for fruitful work on any research project but, as with all collaborative teams, it won't suppress competition and it should not discourage criticism, particularly of the constructive kind — a task that cannot be entrusted to computers, since they are programmed to fulfill orders, not to question them.

All this used to be taken for granted in science studies before the post-modernist onslaught on rationality and objectivity (see Barber 1952; Merton 1973; Worden *et al.* 1975; Zuckerman 1977). However, let us return to an epistemological examination of research projects.

2.3 Analysis of the Concept of a Research Project

The concept of a research project may be analyzed as the following ordered ten-tuple:

$$\Pi = \text{<}Philo, \ Background, \ Problem, \ Domain, \ Method(s), \ Materials, \\ Aim(s), \ Plan, \ Outcome, \ Impact\text{>},$$

where

Philo = The set of philosophical presuppositions, such as the tacit assumption that, as Balzac once said, flowers came before botany;

Background = The body of relevant extant knowledge, such as neuroscience in the case of current scientific psychology;

Problem = The epistemic hole(s) to be filled, such as the nature of dark matter;

Domain = The reference class or universe of discourse, such as faunas in the case of zoology;

Method(s) = The means to be used (e.g., navel gazing, trial-and-error, tinkering, mathematical modeling, measuring, experimenting, computing, or archival search);

Materials = The natural or artificial things to be manipulated (e.g., drugs, animals, plants, or measurement instruments);

Aim(s) = The goal(s) of the envisaged research (e.g., finding a new thing, property or process, formalizing a theory, or testing it);

Plan = A sketch of the course of action, from problem statement to outcome checking to possible impact estimate, such as the benefits deriving from replacing a current polluting industrial process with a "green" or clean process leading to a similar outcome;

Outcome = The output or finding(s) of the investigation, such as a new and more effective drug in the case of pharmacology; and

Impact = The possible effect of the outcome on other projects or even entire disciplines, such as the (unlikely) influence of the study of past economic crises on the design of economic policies.

The following examples may help clarify the preceding sketch:

Example 1: The New Horizons Deep Space Expedition (2006; 2015). Problem: what does the enigmatic Pluto look like at close range? Is it a single body or a system of bodies? Philosophical presuppositions: Pluto and its companions are material things embedded in the solar system, as well as partially knowable through scientific research. Background knowledge: current planetary astronomy, geology, and climatology. Domain: the Pluto six-body system. Methods: (a) calculation of Pluto's and the spacecraft's orbits; (b) astronomical observation and spectroscopic analysis of the atmospheres; and (c) design and construction of a spatial probe due to travel about 2,000 bn km over nine years. Aims: (a) to expand our astronomical knowledge — an item of basic (disinterested) research; and (b) to check the performance of a long-lasting spacecraft equipped with sophisticated yet sturdy instruments — a piece of advanced technological research. Plan: listing the operations, including budgeting, recruiting, and casting the three teams charged with executing them: the technological team dealing with the space probe and its tracking, and the scientific teams devoted to performing and evaluating the observations and measurements.

Example 2: Find out the time window of the anatomical effects on the brain, if any, of the early learning of a complex subject poor in algorithms, such as Euclidean geometry, as opposed to an algorithm-rich subject, such as the infinitesimal calculus. Philosophical presupposition: everything mental is cerebral. Background knowledge: current cognitive neuroscience. Aims: (a) to enrich our knowledge of the traces left on the brain by learning; and (b) to warn education scientists of the lasting effects of negative learning experiences, such as punishment and exclusive reliance on test results. Plan: listing the operations, including budgeting, casting, and recruiting the team charged with executing them.

Example 3: Design a "green" counterpart of one of the polluting and wasteful ("brown") biochemical reactions currently performed in pharmaceutical laboratories. This is a case of both applied-science and technological research, for it seeks new knowledge but it also involves artifacts design — the hub of advanced technology — and it raises economic, health, and environmental concerns.

We have repeatedly asserted that, contrary to the prevailing opinion, scientific research has a number of philosophical presuppositions. They are shown in Figure 2.1.

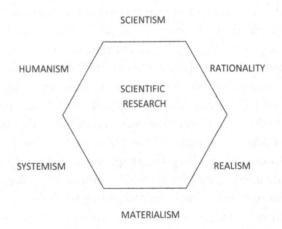

Fig. 2.1. The philosophical matrix of scientific research.

2.4 Research Programs: Successful, Failed, and In-Between

A *research program* may be defined as a sheaf of research projects sharing a theory, philosophy, and goal, and possibly pursued originally in a different discipline. The best-known and most fruitful of the explicit programs in mathematics was the Erlangen program, which aimed at unearthing the algebraic, in particular group-theoretical, foundation of all the geometries. The next successful mathematical program was David Hilbert's effort to enhance formal rigor by rewriting everything possible in set-theoretic terms and axiomatizing all the existing theories.

The famous Bourbaki group took off where Hilbert had left, and it caused quite a stir by its attempt to unveil the bonds among the various branches of mathematics, which had evolved separately from one another, and by emphasizing their set-heoretic foundation. However, Bourbaki was blamed for turning the teaching of mathematics into a boring formalistic exercise deprived of heuristic props. I heard Jean Dieudonné, the school's spokesman around 1960, shouting "*À bas le triangle!*"

Ironically, these champions of rigor all but ignored logic, and their love of novelty was not enough to embrace category theory, which brought the replacement of set theory with category theory as the foundation of mathematics — a move that jibed with Hilbert's injunction to keep deepening the foundations of the discipline. Incidentally, Popper was grateful to his assistant Imre Lakatos "for having proved that mathematics has no foundations." Presumably, the workers in this discipline would retort that math has no *final* foundations — and the same may apply to other sciences.

The earliest research program of factual matters was perhaps ancient atomism, which attempted to explain everything in terms of restless atoms moving in the void. Two millennia later came modern mechanism, from Descartes to the mid-19th century, which tried to explain everything in terms of *figures et mouvements* — as Descartes put it. This program was quite successful, particularly after Newton and Euler, because it was cast in mathematical terms, it suggested a multitude of new experiments and artifacts, and it attracted the best brains in natural science and mathematics.

But mechanism was unable to reduce electromagnetic fields, chemical reactions, and life processes to mechanical processes. Still, after fusing with chemism, mechanism — or rather physico-chemicalism — proved to be far more successful in biology than vitalism. Yet the physiologist and psychologist

Jacques Loeb, who performed experiments to debunk the teleology and the free-will hypotheses, published to great acclaim his *Mechanistic Concept of Life* as late as 1912.

The need to enlarge mechanism became apparent in 1847, when five eminent investigators, headed by the towering Hermann Helmholtz, published the manifesto of the "mechanist" (actually physico-chemicalist) program. The gist of this program was the ontological and methodological thesis that all of nature could and should be accounted for exclusively by physics and chemistry. No *vis vitalis* (life force), *bildende kraft* (constructive force), or even teleology (goal-searching) were to be tolerated in biology, since these spiritualist remains belonged neither in physics nor in chemistry.

The materialism of the five scientists who signed that manifesto — Ernst Brücke, Emile Dubois-Reymond, Hermann Helmholtz, Carl Ludwig, and Friedrich Schwann — was simplified and popularized by the self-styled scientific materialists Ludwig Büchner, Jakob Moleschott, and Carl Vogt. The first three were minor scientists but very popular writers. The manifesto had an immediate and positive impact on biology and related fields, but it was ignored or attacked by the philosophers mesmerized by Kant or Hegel.

Marx and Engels called those writers 'vulgar (or crass) materialists' just because they had no use for Hegel's dialectical "negation," "sublation" (or negation of the negation), or the "unity and struggle of opposites." Marx was so incensed by those non-Hegelian materialists, that he interrupted his work on economic theory to write a whole book against Professor Vogt, who was not only a respected zoologist but also an agent of Napoleon III.

On the other hand Marx praised Ludwig Feuerbach (1947), a minor philosopher who, at a time when German science was growing impetuously, was still wrestling with both theology and Hegel. Later on he announced a "physiological materialism" that never went beyond the programmatic stage. Marx rightly objected that social changes are too fast to be accounted for by biology. Nevertheless, today's Marxists keep the Feuerbach flame while ignoring Holbach, as well as the materialist core of the natural and the biosocial sciences.

Despite Marx's sarcasms, Büchner's 1855 *Kraft und Stoff* (*Force and Matter*) had an instant bookstore success, and remained in print over one century. That popular book may have more for the diffusion of materialism than dialectical materialism, which in the succeeding century became a serious hindrance to the advancement of natural science. Suffice it to recall its support of the charlatan Lysenko and its opposition to the relativity theories, genetics, the synthetic theory of evolution, and cognitive neuroscience.

The great success of the physicochemicalist program in biology is now obvious — if we forget evolutionary biology, which involves the nonphysical concept of history. What is less well known is that their success was largely due not only to its adherence to the scientific method but also to materialism, whereas vitalism had remained attached to backward philosophies like intuitionism and vitalism — a part of spiritualism. Ironically, the prescient, eloquent and fiery *Communist Manifesto* of Marx and Engels was hardly noticed when it appeared one year later.

Yet, ironically, vitalism is still alive, e.g., in the popular thesis that all life processes are goal-directed, despite the cybernetic hypothesis, fathered by Norbert Wiener and his associates, that the appearance of teleology is just a result of negative feedback or control devices in organisms or artifacts. Wiener failed to recognize this explanation as a triumph of materialism because he believed that the concept of matter had been superseded by that of energy — as if energy could be anything other than a property of a material thing θ, as is obvious when writing, for example, $E(\theta, f, u)$ = real number, for the energy of thing q, relative to frame f, and measured in unit u (e.g., electron-volt).

The next grand and successful scientific program was Darwinism, which rose in 1859 and inspired a whole new worldview. This program was greatly strengthened in the 1930s by the synthesis of evolutionary ideas with genetics, and it culminated in 1953 with the Crick and Watson discovery of the DNA structure, which kicked off molecular biology.

Inevitably, this triumph of chemicalism gave rise to genetic determinism, the thesis that "genome is destiny." The popularizer Richard Dawkins exploited the Crick–Watson revolution with his "selfish-gene" fantasy (1976), which involves the false assumptions that what are selected are not whole organisms but their genomes: that the organism is just a gene funnel between generations [of what?], and the environment only a sieve.

The idea that "nature" (heredity) trumps "nurture" (environment) is still being discussed in some circles despite rendering the very existence of the organism "paradoxical," as Dawkins himself put it. The recent emergence of epigenetics, which has shown the vulnerability of the genome to environmental stimuli, and even the heritability of some chemical alterations in the genetic material, is the last nail on the coffin of geneticism, but philosophers have yet to notice it.

However, evolutionary biology has kept growing and fusing with other biological disciplines, chiefly ecology (eco-evo), developmental biology (evo-devo), and paleoanthropology. Dawkins also was complicit with the failed program of pop evolutionary psychology, the successor of sociobiology — Edward O. Wilson's brilliant if flawed brainchild. Both of these projects have tried to explain everything social — selfishness and altruism, monogamy and poligamy, racism and language, free market and monopoly, cooperation and war, religion and science — as useful adaptations. However, the evolutionary psychologists added the proviso that man has not evolved since the end of the Pleistocene, so that we would be but "walking fossils" (Buss 2015).

The idea of putting these fantasies to the test did not occur to the inventors of these modern myths, whence they were not scientific projects (Pievani 2014). It would seem that every revolution, whether political or cultural, provokes a counter-revolution.

Another outstanding scientific program — though this time a successful one — is the attempt to fulfill the Hippocratic conjecture that everything mental is cerebral rather than spiritual. Starting with the publication of Donald Hebb's seminal *Organization of Behavior* (1949), nearly all the important findings in psychology, in particular the localization and interdependence of a number of mental processes — from perception to invention, and from anxiety to moral dilemmas — have been so many successes of the psychoneural identity program.

By contrast, information-processing psychology, which stems from the opinion that the brain is a computer, has not explained any mental processes, nor it has it guided the design of any therapy to treat mental diseases. A major cause of the failure of the digital program in psychology is its utter disregard for brain science, which has shown that the brain of the upper vertebrate has functions, such as self-activation and lateral inhibition, which the other bodily organs lack. In short, digital psychology is a failed program.

This failure should have been foreseen from the start, because the digital research program did not include experiments and it violated the external consistency condition, in this case compatibility with brain science. But the program was attractive in being materialist and simple.

Lastly, there is the rational-choice program in social studies, which has been mainstream dogma since about 1870. The chief hypotheses of the various rational-choice theories, from standard microeconomics to ethics, is that everything social results from free individual decisions, and that these are guided by the wish to maximize the actor's expected subjective utility or benefit — the product of the utility of an action by its probability. The upholders of these ideas have ignored the many criticisms directed at them, from the lack of rigorous definitions of the key notions of subjective or personal probability and utility, to its predictive poverty and lack of experimental control.

In short, the rational-choice and genetic-determinism programs continue to be popular despite having failed. Ditto string theory, the multiverse fantasy, and the it-from-bits extravaganza. What is common to all of these failed research programs is that they lack experimental support and violate the external consistency condition. Shorter: their failure is due to the fact that they were not sheaves of *scientific* research projects. Ironically, Freud's early program of a biological psychology was scientific, but he did not even attempt to work on it. His defenders keep mentioning this failure in an attempt to prove that Freud was basically a scientist after all. Contrary to a rather popular opinion, psychoanalysis is not a failed science, because it did never involve scientific research. But there is no denying that it had a tremendous if negative impact on clinical psychology, pop social studies, and literature.

Finally, the *Annales* school, which flourished in France between ca. 1930 and ca. 1990, was the deepest, most ambitious, and most successful historiographic program since Ibn Khaldûn's in the 14th century. Indeed, the *Annales*' aim was no less than an attempt to substitute *l'histoire totale* for the arid description of isolated political, military and dynastic events that missed all the slow but radical (*longue durée*) social changes. Although this school dissolved almost overnight along with the Soviet bloc in 1991, its systemic philosophy, which Fernand Braudel put into practice in his monumental *La Méditérranée* (1949), is here to stay along with the economic globalization announced by Marx in 1848. Fortunately, the Spanish historian Josep Fontana (2011) continued the work of the *Annales*.

2.5 Science: Authentic and Bogus

Experienced scientific investigators in good faith take it for granted that the research project they are about to evaluate or tackle is authentically scientific rather than one in bogus science. In fact, scientific frauds are infrequent, and nowadays mostly confined to the biomedical sciences — perhaps because medical students do not train as scientists, and their mentors are under the "publish or perish" pressure.

Let us briefly recall three of the most highly publicized recent frauds. One of them was the claim by Jacques Benveniste and his team, in 1988, that homeopathy works because water remembers the active ingredients that had formerly been dissolved in it. A year later Martin Fleischmann and Stanley Pons announced that they had achieved cold nuclear fusion in a kitchen. Both turned out to be cases of self-deceit and junk science. What can be said of the hundred or so laboratories around the world that annunced that they had successfully replicated the Fleischmann–Pons fiasco?

In contrast, Marc Hauser, a co-author of Noam Chomsky's, commmitted a deliberate fraud. In fact, he was forced to resign his Harvard professorship in 2014 when he admitted having knowingly altered his findings on the biology of cognition and morality in non-human primates — which purported to support Chomsky's claims that primate vocalization is impervious to social feedback, and that human language is unrelated to nonhuman primate communication. (See Takahashi *et al.* 2015 for relevant experimental findings about marmoset vocalization.)

In view of our antecedent knowledge, the homeopathy (or water memory) finding *had* to be bogus for two reasons: (a) because the dilutions were of the order of one in 10^{120} molecules, which amounts to one molecule per galaxy — an inefficacy guarantee; and (b) because no plausible mechanism for such memory was proposed, let alone found. Both reasons place the fantasy in question beyond the fringe of science.

Both the water memory and the cold fusion flops seem to have derived from ignorance of the scientific method rather than from bad faith. After all, such ignorance is pervasive in the science community. In fact, most scientists do not think critically and productively all of the time, but apply certain standard techniques, such as microscopy, brain imaging, or computer programming. Most of them are satisfied if they obtain clearer images or more precise numerical values. Only Santiago Ramón y Cajal had the temerity to guess — alas, correctly — the direction of nerve currents from looking at the dead neural networks that Golgi's staining technique had revealed.

In conclusion, these are the main features of the pseudosciences:

a. They are outdated and do not conduct scientific research, so that they never come up with new authentic knowledge — by contrast to the sciences, which produce new items every week.

b. They are the property of cliques whose members never publish in refereed journals or meet in open congresses.

c. They can be learned in a short time, for they always have the same answers to the same questions, whereas learning any science to the point of being accepted as a member of a scientific community takes many years of disciplined and laborious apprenticeship.

d. They have false or even extravagant philosophical presuppositions, such as the "power of mind over matter."

e. They are dead ends instead of suggesting new problems or methods.

f. Most of them are practiced for profit rather than for the benefit of humankind.

g. Some of them are harmful to individuals — as is the case of the alternative medicines — or even entire peoples, as happens with the outdated and antisocial dogmas of economic orthodoxy, which justify social policies that impoverish entire peoples.

2.6 The Received View of Science

We are finally in a position to attack the central target of the present book, namely the vulgar view of scientific work. This is the belief that scientific research is just a combination of common sense with rigorous logic, meticulous observation or computation, and honest reporting; that it is always data-driven, never curiosity nor hunch-driven; that it has no philosophical underpinnings; and that its results may be condensed into simple formulas, neat diagrams, or succint technical reports.

Ptolemy (2nd century C.E.), perhaps the greatest astronomer of antiquity and a formidable critic of the heliocentric hypothesis, emphasized that the entire scientific exercise, from the object under observation to the final hypothesis, had to be confined to phenomena, that is, appearances, since anything beyond them had to be speculative, hence beyond science. Shorter: his slogan was *Stick to appearances* (see Duhem 1908).

This view is of course *phenomenalism*, a synonym of radical empiricism — the thesis that knowing is basically experiencing. This is the vulgar epistemology, as well as the theory of knowledge of such celebrities as Hume, Kant, Comte, Mach, the logical positivists, and the faithful of the Copenhagen interpretation of the quantum theory — who claim to deal only with observables.

In stating that the universe is the collection of phenomena (ontological phenomenalism), or that these alone are knowable (epistemological phenomenalism), a consistent phenomenalist cannot understand the warning often affixed on the rear-view mirrors of cars: "The object on the mirror is closer than it appears." Thus, phenomenalism is hazardous on top of being false.

Nor can phenomenalists understand why neuroscientists and psychologists spend so much time studying the organ that transforms sensory inputs into phenomena, namely the brain. Nor can they grasp why the realists, from Galileo and Descartes onwards, emphasized the phenomenal/real, and the qualia/primary properties distinctions.

In addition, phenomenalism is anthropocentric, for it professes to deal only in appearances, that is, facts that happen only inside human brains. But these organs exist only in a tiny region of the world and, furthermore, they are latecomers. Hence Kant's assertion, that the world is the sum-total of appearances, has been falsified by evolutionary biology, according to which sentient beings emerged only about two and a half billion years ago.

Yet phenomenalism is still alive and well in the physics community, where reference frames are often wrongly identified with observers, and consequently "frame-dependent" is confused with "observer-dependent," and where dynamical variables are called "observables" — as if observability could be decreed and, moreover, as if it were an intrinsic feature of factual items rather than a relational one.

That the latter is the case is suggested by the fact that, if pressed, any experimentalist will admit that the expression 'Variable V is observable' is short for 'Property V is observable by such and such means under such and such conditions.' The means in question are not just instruments such as telescopes, but also indicators, such as the pointers of a measuring instrument or the tracks left on mud by an organism that lived millions of years ago. Only an evolutionist can "see" (conjecture) a fossil in the old bone that the farmer's dog dug up.

In the study of mind and society, phenomenalism is called *behaviorism*, for it urges students to confine themselves to observing the outward behavior of their subjects, without speculating about their mental processes, such as feelings, intentions, motivations, valuations, decisions, and plans. Undoubtedly, behaviorism yielded some valuable findings when practiced by rigorous and imaginative researchers such as Burrhus Skinner, but most of its findings have at best posed interesting problems about the brain mechanisms producing observable behavior.

At worst, the observation of outward behavior by individuals lacking in scientific training has produced works like *Laboratory Life*, by Bruno Latour and Steven Woolgar (1979), who spent a year observing a team of molecular biologists. Since they did not understand the scientific problems their subjects were grappling with, they confined themselves to taking note of such trivia as making observations, taking pictures, and exchanging gossips during coffee breaks.

To make sense of such trivial observations, Latour and Woolgar made up the ridiculous stories that doing science boils down to making inscriptions and talking shop in hopes of gaining power, and that in the process "scientists construct facts" rather than just studying them. As if this were not enough, their book gained their authors instant "bookstore celebrity." This fame was well deserved: they had reheated constructionism–relativism, an essential component of both old subjective idealism and fashionable postmodernism, and one so simplistic that it was accessible to everybody. (See further criticisms in Bunge 1999a, 2011; Sokal & Bricmont 1997.)

2.7 Phenomenalism Hinders the Advancement of Knowledge

Phenomenalism hinders the advancement of knowledge because it advocates a radical shrinking of the domain of facts and properties, namely those that are observer-dependent, whereas genuine science is impersonal. To realize this point, suffice it to recall a handful of landmarks in the history of science.

2.7.1 *Heliocentrism*

Observation shows the world, in particular our star, to orbit around our planet. Whoever adopts Ptolemy's advice, to stick to appearances, must adopt the geocentric model of the solar system, and consequently s/he will reject the heliocentric model, usually attributed to Philolaos of Kroton (5th century C.E.) and Aristarchos of Samos (3rd century C.E.). This hypothesis remained buried, and it was rejected by the few who knew of it, until rescued much later by Copernicus, Galileo, Kepler, Huyghens, and Newton.

Newton was the first to explain why the planets *had* to circle the sun rather than the other way round: because of their far smaller masses, they are dominated by the sun's gravity. The astronomical observations available around 1600 were not precise enough to reject the geocentric model, on top of which it agreed with common sense as well as with the book of Genesis. Hence the Galileo trial.

The last doubts about heliocentrism were swept away only around 1700. But phenomenalism, chased from science, took refuge in George Berkeley's philosophy (1710), whose radical phenomenalism was embraced by Hume (1748) and Kant (1787), who in turn inspired positivism from Comte to Mach to the Vienna Circle (mid-1920s–1938) and the Unified Science movement (1938–1962).

Incidentally, two prominent members of this interesting movement, Philipp Frank and Hans Reichenbach, claimed that, since motion is relative, the two models in question are mutually equivalent. They ignored the fact that ours is just one of eight planets subject to the sun's gravitational field, whose intensity makes it the dominant member of the planetary system. In addition, both men failed to realize that, because of its phenomenalism, positivism is subject-centered, hence naturally sympathetic to geocentric. Nor did they realize that in this respect they belonged in the same party as their archenemy, Husserl's phenomenology (or egology).

Roughly the same subjectivism was revived in the 1960s and 1970s by the constructivist-relativists who hijacked the sociology of science in the wake of the Kuhn–Feyebend coup. Those newcomers claimed that scientists did not discover anything, but invent their objects, and morever do so to increase their "social capital" or even their political clout (Latour 1987).

2.7.2 *Atomism*

Ancient atomism, the worldview invented by Leucippus and Democritus, perfected by Epicurus, and sung by Lucretius, was yet another triumph of serious materialist speculation — as well as the kernel of a rather popular secular ideology. Although it lacked empirical evidence, ancient atomism explained the rndom motions of dust particles in a sunbeam, as well as some imperceptible processes, such as the drying of wet clothes and sails exposed to sunlight, that its rivals left in the dark.

Greek atomism attracted some of the early moderns, such as Thomas Hobbes, because it outshone the occult powers of the schoolmen. But, because of its lack of empirical support, the theory remained outside science until Daniel Bernoulli rescued it in 1738 to build his kinetic theory of gases. Around 1810 John Dalton, and independently Jöns Jakob Berzelius, Amedeo Avogadro, and later on Stanislao Cannizzaro, proposed atomic explanations of chemical compositions and chemical reactions — at about the same time that Hegel, Fichte and Schelling were spewing their extravagant *Naturphilosophien*. (See Bunge 1944 for a criticism of the philosophy of nature.)

Ironically, Dalton thought that his own greatest achievement was his meteorological diary, which he kept for 57 years in hopes of finding a meteorological law, in accordance with the ruling inductivist methodology. But of course Dalton is honored not for his meticulous weather observations, which led nowhere, but for his bold atomic conjectures, such as his formula for the composition of water, namely HO — not a bad first approximation to the true formula H_2O.

Yet physicists remained skeptical about atoms even after 1871, when Ludwig Boltzmann reduced thermodynamics to statistical mechanics. In 1827 the botanist Robert Brown had observed the random motions of pollen grains in water. But only Jean Perrin's experimental work on them in 1908 persuaded most physicists that those jerky movements were due to the random impacts of water molecules on the pollen grains. His work was made possible by Einstein's theoretical paper published three years earlier, for it told experimentalists *what* to measure, namely the mean squared displacement of a particle over a given time interval.

From then on till our days, the progress of atomism has been as relent-less as dizzying. Suffice it to recall its most popular early successes: the analysis of radioactive products into alpha-rays, electrons, and gamma rays; Rutherford's pioneering experiments in nuclear physics, which motivated and confirmed the Rutherford–Bohr planetary model of the atom; and the explanation of sunlight as a byproduct of the synthesis of four hydrogen atoms into one helium atom — an example of the relativistic formula "emitted energy = mass defect $\times c^2$." All of these findings reinforced realism and refuted the vulgar view of science, that scientific theories are data concentrates.

2.7.3 *Biological evolution*

The ancient materialist philosophers Anaximander (6th century C.E.) and Empedocles (5th century C.E.) were perhaps the first to suspect that all living things evolve and, in particular, that humans descend from fish. Although at the time this was but a hunch, it was no wild fantasy, for it explained facts such as the finding of fossilized remains of marine animals on mountain tops: maybe these had been under the water long ago, before geological upheavals pushed them upwards.

Buffon, Diderot, Maillet, Maupertuis, Erasmus Darwin (Charles's grandfather), and a few others embraced evolutionism long before the detailed study of fossils, later on of fruit flies, and finally of bacteria and viruses as well, turned evolutionary biology into an experimental sci-ence and fused it in the 1930s with genetics (the synthetic theory of evolution), and around 2000 with developmental biology (the evo–devo synthesis).

Once again, hypothesis preceded and guided observation. In particular, the evolution hypothesis generated paleontology, as it encouraged fossil collectors to see their findings as remains of organisms. In fact, before that, fossils had been regarded as stones, and therefore as the property of geolo-gists. And some believed that those peculiar stones were the product of jokes played by nature. In any event, paleontology was not born from hypothesis-free observation. And it might not have been born had the Scientific Revolution not revived Heraclitus's philosophical principle *panta rhei* (everything flows).

2.7.4 Quanta: from observation to quantum theory and back

In 1814 Joseph Fraunhofer saw and described for the first time dark lines in the solar spectrum. Many other investigators followed, and studied the spectra generated by the electrical excitation of a large number of substances. Eventually some patterns emerged, such as Balmer's series in the hydrogen spectrum (1885), and the corresponding semi-empirical equations were formulated.

By 1850 spectroscopy was in full bloom, but it took six decades for the mechanism behind the spectral lines to be unveiled. The earliest successful theory was Bohr's theory of atoms as miniplanetary systems in 1913, which explained each spectral line as the effect of the transition of the corresponding atom from one energy level to a lower one. Incidentally, energy levels are unobservable: only the light emitted or absorbed by a transition between such levels is observable.

The limits of Bohr's theory were soon recognized, but it took another decade for the so-called classical quantum theory, with its iconic elliptical orbits, to be superseded by modern quantum mechanics, built chiefly by Heisenberg, Schrödinger, Born, Pauli, Dirac, and de Broglie. Although some of these physicists claimed to stick to phenomena, hence to dispense with unobservables, the truth is that the typical variables of quantum mechanics — in particular the energy operator and its eigenstates, as well as the corresponding state (or wave) function, are unobservable (see, e.g., Bunge 1967a, 1973a).

Eventually Heisenberg admitted this philosophical "flaw," and in 1937 attempted to correct it with his S-matrix, believed to summarize scattering experiments, that is, to contain only variables referring to incoming and outgoing particle beam colliding with a target. This theory made some noise in the early 1960s, but turned out to be impotent without the help of the quantum theory. Later inventions, such as quantum field theory and chromodynamics, were further steps away from observation. In particular, the trajectories of electrons and photons occurring in the Feynman diagrams are semi-classical fictions helpful only to perform certain calculations (Bunge 1955).

Furthermore, quarks, assumed to be the elementary constituents of protons and neutrons, do no exist in isolation but only in combination, so that they are unobservable in principle. Furthermore, it is currently estimated that four-fifths of the matter in the universe is "dark," that is, invisible, for it neither emits nor reflects radiation.

In short, the basic constituents of observable lumps of matter are unobservable. Hence observability implies reality but the converse is false, so that *Real ≠ Observable*. This result strongly vindicates ancient atomism. By the same token, it completes the downfall of phenomenalism in physics, though it survived in philosophy. This is why it has been said that philosophy is where science goes to die.

Astrophysics and cosmology yield a similar result: some of their referents too are hardly observable while others, like Neptune and black holes, have been looked for only because certain calculations indicated that they should exist. In particular, the existence of black holes is confirmed by the observation of stars that seem to revolve around an empty point (see Romero & Vila 2014).

In short, observation alone cannot suggest deep and general theories, which Einstein regarded as "free creations of the human mind." But the matter of their truth is quite different from that of their genesis: only observation or experiment can provide some evidence for or against the truth-value of a hypothesis or theory concerning a parcel of reality. But again, no empirical procedure will be designed unless the idea to be tested is deemed to be scientific. Let us therefore tackle the scientificity issue.

2.8 Scientificity: Demarcation Problem

Nobody has the time and resources required to check in detail all the projects submitted to a scientific publication or a funding agency, whence the need for a preliminary screening. I submit that such a filtering consists in determining whether or not the project in question is scientific long before its results can be pronounced to be at least approximately true in the case of basic or applied science, and feasible as well as useful in that of technology.

I propose that a research project

Π = <*Philo, Background, Problem(s), Domain, Method(s), Materials, Aim(s), Plan, Output, Impact*> is *scientific* if and only if

a. its philosophy is *realist* or objectivist (rather than either radically skeptic, subjectivist, or conventionalist), *materialist* (rather than spiritualist), and *systemic* (rather than either holistic or atomistic);

b. its background is *up to date and in flow* rather than outdated and ossified;

c. the problems at hand are well posed, and neither trivial nor over-ambitious;

d. its domain is partially known or suspected to be *real and accessible* rather than being esconced in a parallel universe disconnected from ours;

e. its methods are *scrutable, impersonal,* and *perfectible,* and among them impersonal observation, tinkering, controlled experiment and computer simulation stand out;

f. its chief aims are finding new truths in the case of basic science, and novel artifacts of possible practical utility in the cases of applied science and technology;

g. its research plan can be *implemented* with the envisaged means;

h. its means and results are *reproducible* by other similarly equipped researchers; and

i. the solution to its central problem(s) is likely to constitute a *valuable contribution* to our knowledge or our welfare rather than either trivial or an excuse for intellectual gymnastics like chess. (See Bunge 2003a for the philosophical terms occurring above.)

The preceding stipulation or convention may be clarified by the consideration of the dual concept of an *unscientific* project, that is, one meeting the following conditions, typical of both the nativist and the information-processing psychologies, as well as of the interpretivist (*verstehende*), the rational-choice, and the constructivist-relativist speculations in social studies:

a. Its philosophy is irrealist, in particular subjectivist (subject-centered), as well as spiritualist ("mind over matter"), and either holist ("the whole precedes its parts") or individualist ("there are only individuals") rather than systemic ("every thing is a system or part of one");

b. Its background is dated;

c. Its problems are ill posed, or working on them requires knowledge that the grant applicant(s) lacks;

d. Its domain is not known to be real and accessible;
e. Some of its methods are inscrutable or ill-designed — for example, its experiments are not such because they do no involve control groups; and the use of Bayesian (personal, subjective) probabilities makes one doubt of any project that makes intensive use of them;
f. Its plan is unfeasible, at least with the resources in hand; and
g. Its aim(s) are merely rhetorical, or worthy but unattainable by the investigator(s) in real time.

No doubt, sometimes our stringent objective criteria are not met, either inadvertently or intentionally. But eventually the resulting fault is likely to be found out — which is why scientific research is often characterized as *the self-correcting intellectual process*.

The problem of the scientific worth of a theory or an empirical procedure recurs in the courts of law every time expert witnesses intervene. In fact, judges are expected to decide whether the testimonies of such witnesses are scientifically valid or just opinions.

Sometimes the experts do not concur on what makes a testimony scientifically reliable. In such cases they could use what the Chief Justice of the UK calls a 'scientific-method primer' (Neuberger 2016). Regretfully, no such primer is likely to be produced by philosophers, divided as they are on what makes an item scientific. Let us therefore see how the scientific communities evaluate research results.

CHAPTER 3

EVALUATION OF RESULTS

So far we have discussed the reasons for surmising that a given research proposal is worth pursuing. Supposing that its results are in, how shall we evaluate them? Only such evaluation may tell whether the effort was worth doing — supposing that the outcome is definite rather than inconclusive.

3.1 Success Criterion: New Truths

Since about 1600, it has generally been accepted that a given research project in factual science has been successful if and only if it has yielded *new truths*, as warranted by observation or experiment, as well as by external consistency, or compatibility with the bulk of antecedent knowledge. An old academic joke puts it like this: that piece of work is worthless because, while its true results are not new, its novel results are false.

At all events, empirical confirmation has usually been regarded as both necessary and sufficient for truth, and empirical refutation for falsity. Even Karl Popper, who boasted of having slain the empiricist dragon, fell for that criterion, which is clearly empiricist because it exalts experience as the supreme arbiter, on top of which it demands full truth, while in practice one usually settles for partial truth.

In addition, Popper's claim, that only falsifiability matters, is logically flawed, for the refutation of a proposition A is logically equivalent to the corroboration of not-A. Actually, the alleged asymmetry between refutation and confirmation boils down to framing, which is often a rhetorical device rather than a methodological move.

For example, most people have been more impressed by the finding that the mental is cerebral, than by the refutation of the myth of the immaterial mind. Thus, even the great Francis Crick, writing in 1994, found "astonishing" the hypothesis that everyting mental is cerebral, alhough it had been formulated by Alcmaeon (ca. 500 C.E.), adopted by Hippocrates and Galen, and placed at the center of biological psychology since Paul Broca and Karl Wernicke created it in the mid-19th century.

In any event, ordinarily scientists are expected to explore a part of reality and to unveil some of it. This is why the vast majority of Nobel Prizes in science have been awarded for *discoveries* or *findings* — in other words, for new factual truths, i.e., truths about matters of fact, such as Hans Selye's uncovering of stress as a major source of disease, James Olds and Peter Milner's accidental discovery of the reward (pleasure) center, Wilder Penfield's somatosensory map on the cerebral cortex (the so-called homunculus), and John O'Keefe's discovery of the positioning system in the mammalian hippocampus.

No Nobel prizes have ever awarded for mere refutations. One reason is that doubt and denial are unproductive and far cheaper than well-grounded assertion. For example, the denial that the earth is flat is consistent with the assertions that it spherical, ellipsoidal, pear-shaped, conical, cylindrical, etc. This is why, in proving that the earth is round, Magellan's voyage around the earth deserves a much larger credit than any of the many sailors who had confided their skepticism about the prevailing flat-earth dogma to their audiences in European taverns around the year 1500, but had not ventured to suggest the precise shape of our abode.

If a research project does not aim at finding previously unknown facts or truths about them, it does not deserve to be called *scientific*. This is of course a truism about truth of the factual kind, but it bears repetition given the postmodernist disregard for truth.

To find new truths about reality we must make observations or experiments because the universe has pre-existed humans — a realist philosophical presupposition. True, powerful theories may anticipate certain events — but only if conjoined with the relevant empirical data. No data, no factual truths, hence no factual science. Why is this so? Because of the definition of factual truth as adequacy of a proposition to the facts it refers to. (The notion of truth as idea/fact matching can be rendered precise: see Bunge 2012b.)

And only observation, and even more so experiment, which involves the comparison between the experimental and the control groups, can tell us whether the facts in question really occur or are figments of our imagination. This is so because only such intervention puts us in close touch with reality. Indeed, the simplest *reality criterion* is this: For all x, x is real if and only if the existence or occurrence of x makes a difference to something else, preferably an artificial detector or a measuring instrument.

This is why thousands of physicists tried for decades to detect gravitational waves. Most of them believed in their existence just because Einstein predicted them in 1915 as part of his theory of gravitation, which also predicted the existence of some thirty other "effects," among them the existence of gravitational lenses and black holes. But the actual discovery of the said waves came only in 2015.

In other words, the hypothesis of the existence of the elusive gravitational waves was scientific from birth, not because it is falsifiable but because it is precise, it coheres (or fits in with) a generally accepted theory, and because failure to detect the waves in question could be blamed on their extremely weak energy, which in turn is attributed to their being ripples in spacetime rather than "particles."

The LIGO team, the first to detect them — that is, to corroborate Einstein's hypothesis of their existence — is sure to earn a Nobel Prize, while the timid folks who just noted its falsifiability can only congratulate the thousand ingenious folks who delivered the goods at the cost of about US$1.1 billion. Incidentally, which private enterprise would have paid this bill for a discovery with no foreseeable uses?

The history of the neutrino is parallel, though even more important and dramatic. About 1930, an anomaly was noted about beta decay. This process consists in the transmutation of an atom together with the emission of an electron, as in carbon-14 $\rightarrow$ nitrogen -14 + electron. The energy of the nitrogen plus the electron seemed to be smaller than that of the parent. To preserve energy conservation, Wolfgang Pauli conjectured that an additional particle was emitted. This newcomer was called *neutrino* because it was electrically neutral, as well as massless or nearly so, and therefore very hard to detect.

An ingenious and huge detector was designed, and in 1995 neutrinos were finally discovered — 65 years after having been conceived. Later on, neutrinos were used to turn protons into neutrons plus positrons. Further related discoveries — for instance, that there are two kinds of neutrinos, and that both cosmic rays showers and solar radiation contain neutrinos galore — came as a bonus. In all of these cases, unexpected observations were made in an effort to *conserve* a crucial theoretical principle, not to start a revolution, as Kuhn might put it, or to refute a myth, as Popper would claim.

3.2 Falsifying Falsifiabilism

In 1935 Karl Popper caused consternation in the pro-science camp by stating that, while scientific theories cannot be proved, they can be disproved, and furthermore, that only falsifiability makes them scientific. In other words, he claimed that if an idea cannot be falsified at least in principle, then it is not scientific but pseudoscientific or ideological. Hence he recommended scientists to attempt to falsify their own conjectures rather than try and corroborate them.

Although Popper's scientificity criterion, namely falsifiability, has become quite popular, it has been argued that it is all-around false — logically, methodologically, psychologically, and historically (Bunge 1959c, Gordin 2015). To begin with, Popper's use of the word 'theory' was careless, for he intended it to cover both a hypothesis and a theory proper, i.e., a hypothetic-deductive system. This point is important because, while a single hypothesis may be either confirmed or refuted by a crucial experiment, the same cannot be ascertained about a theory, for it is an infinite set of statements. In this case one tests only a few important members of the set, and hopes that the rest would turn out to have the same truth-value.

Second, in classical logic, which is the one used in science, the proposition "*p* is false" is equivalent to "not-*p* is true." Hence, confirmation is not weaker than falsification. On the contrary, *negation is weaker than assertion*, because finding that *p* is false is compatible with infinitely many alternatives to *p*. Thus, finding that Aristotle erred in holding that the heart is the organ of the mind gave a chance to other organs, such as the spleen (as the traditional Chinese medics believed), the pineal gland (as Descartes conjectured), and the brain (as cognitive neuroscientists have established). This is why naysayers are much more numerous than the rest of us, who sometimes risk our reputations by making assertions with insufficient evidence.

Third, the sentence 'p is testable' is incomplete, for testability is relative to the test means. For instance, the ancient atomists lacked the detectors and other laboratory instruments required to test their conjectures. In short, the predicate 'testable' is binary, not unary, so that the sentences of the form 'p is testable' should be completed to read 'p is testable with means m.'

Fourth, nearly all scientific empirical observations, measurements, or experiments are performed to *find* something, seldom to falsify a conjecture. If in doubt, look at any citation for a Nobel Prize in natural science. For example, several astronomical observatories are currently working to find the ninth planet in our solar system, which theoretical astronomers have predicted. Since Planet 9, though supposedly gigantic, is gaseous and even more distant than Pluto, the project is thought to require extremely sensitive detectors, and to take at least five years. So, let us stay tuned and prepared to end up by repeating that Planet 9 has *so far* escaped detection. In other words, as long as some astronomers work on the Planet 9 research project, the naysayers will have to keep silent, whereas the optimists can keep up their hopes.

Such hopes of satisfying their curiosity are what make scientists tick. Only masochists and psychopaths work to cause pain to self or others. In short, Popper's advice, to try and topple one's favorite guesses, is psychologically false in addition to being logically and metodologically flawed.

Finally, falsifiabilism is historically wrong. Indeed, most of the myths that turned out to be false were eminently falsifiable to begin with. Let us refresh our memories.

Example 1: Augustine refuted astrology by inventing the story of the two babies born at the same time and in the same household, hence "under the same stars," but one free and the other slave — and yet with hugely different life stories.

Example 2: The four-element theory, held for nearly two milllennia, was refuted by the 19th-century chemists who discovered or made previously unknown authentic elements. In the 1860s, when Dmitri Mendeléyev published his periodic table, 63 elements were known. Today we know nearly twice as many, and counting. The latest to be synthesized is #118, temporarily called *Uuo*.

Example 3: Palmistry, homeopathy, acupunture, parapsychology, psychoanalysis, and spiritual healing were falsifiable from the start, but only a few thought that they were unscientific for failing to propose verisimilar mechanisms for their alleged successes.

The case of psychoanalysis is similar though more complicated. Although the Oedipus story is indeed irrefutable when conjoined with the represssion myth, the remaining psychonalytic hypotheses — in particular the ones about infantile sexuality (before the emergence of sex hormones!), the anal/oral personalities, and social protest as a case of rebellion against the father figure — were falsifiable, and have been abundantly falsified by experimental psychologists. Psychoanalysis was never scientific because neither Freud nor his followers ever did any scientific research. In short, Popper's scientificity criterion does not work. This explains why he approved of steady-state cosmology and standard economic theory, and embraced psychoneural dualism, while regarding evolutionary biology as "a metaphysicsl research program."

3.3 Empirical Corroboration Is Not Enough

Experience is not the unappealable arbiter in thought/fact matches because, when data seem to impugn a fruitful hypothesis, many a scientist will rush to save it by proposing a plausible *ad-hoc* hypothesis. For example, when people objected that they did not perceive evolution in multicellular organisms, Charles Darwin resorted to the incompleteness of the fossil record and the minuteness of intergenerational changes.

Donald Hebb (1951), the founder of contemporary cognitive neuroscience, wrote unfazed that "If apparent contradictions of a useful law are observed, one promptly postulates something else to account for them instead of discarding the law." But, of course, that something else, the *ad-hoc* hypothesis framed to save one's favorite, has got to be independently testable (Bunge 1973b). For example, when certain measurements seemed to falsify Einstein's special relativity, it was conjectured that a vacuum leak in the apparatus was to blame — as it became apparent a few years later.

By contrast, Freud's repression fantasy, designed to protect his Oedipus myth, was not tested independently — and found to be false — until much later: it was a bad-faith ad-hoc hypothesis. Whereas in Einstein's case systemicity was used to protect a truth, in Freud's case it protected a myth. In short, ad-hocness can be in good on in bad faith.

Empirical corroboration is a necessary but insufficient truth indicator. A further truth condition is what may be called *external consistency* (Bunge 1967b). This is the requirement that the new idea or procedure be *compatible with the bulk of antecedent knowledge* — obviously not all of it, since the finding in question is expected to be new in some respect. Hilbert (1935: 151) was perhaps the first to explicitly demand consistency with neighboring disciplines. Let us examine a couple of famous cases.

Maxwell assumed the existence of displacement currents in the dielectrics between the terminals of a capacitor long before they were found experimentally, and he did that solely to save the hypothesis that the total electric charge in an electric circuit was constant. But this ad-hoc hypothesis was testable, and it did not contradict the little that was then known about insulating materials.

The external consistency requirement holds, with all the more reason, for much grander conjectures. For example, the ideas of creation of matter or energy out of nothing, as well as those of telepathy and precognition, of untrammelled economic and political freedom, and of equality without freedom and solidarity, violate that condition. Ditto the assumption in string theory, that physical space is not three-dimensional but 10-dimensional. Consequently, any research project involving them should be deemed to be groundless. And yet string theory and its relatives have dominated theoretical particle physics for a quarter of a century despite that serious flaw and without any experimental support — which says something about the gullibiliy of the individuals concerned.

Third and last, the set of basic research problems should be regarded as public property. In addition, to be fruitful, work on any of them should be neither guided nor constrained by political motivations, for — contrary to Michel Foucault and his followers — cientific research and controversy are about truth, not power (see Raynaud 2015).

The self-styled libertarians urge the privatization of everything, even of science. They do not know that science started to advance at high speed only in the 19th century, when amateurs were replaced by professionals, and universities became public. The massive privatization of universities would kill basic science, because business firms have no use for pure mathematics, particle physics, astrophysics, evolutionary biology, anthropology, archaeology, historiography, and similar projects that are cultivated just to satisfy curiosity.

When science is privatized, the scientific project turns at best into a technological adventure, without regard for either morality or the public interest (see Mirowski 2011). For example, some private pharmaceutical companies have patented many of our genes, so that we no longer fully own ourselves (Koepsell 2009). And some universities are currently trying to shift their professors from papers to patents. Fortunately, others are working against this trend, and towards a free access policy. For example, the exemplary Montréal Neurological Institute and Hospital is refusing to patent any of the discoveries of their researchers.

3.4 Scientificity Indicators

When a scientific research project is submitted to a granting agency or to a scientific publication, it is evaluated by a panel of judges, most of whom have never lost any sleep over the scientificity issue: they rely on their own experience, the candidate's track record, and the interest, promise, and feasibility of the project.

Such an intuitive peer review usually worked well in the traditional fields, and as long as the judges were impervious to political pressures. But in other cases the procedure has been flawed. Witness some of the grants given to projects involving wild speculations in particle physics, cosmology, psychology, and the social sciences — not to mention plagiarism and its dual, the rejection of good papers out of misunderstanding or turf protection.

These failures of the peer review procedure have grown so fast in recent times, that their scientific study has become a new research field with its own journal, *Research Integrity and Peer Review* (2016).

Such waste of public funds suggests observing explicit and well-grounded evaluation criteria. I propose the following *battery of scientificity indicators* to be met by project purporting to be scientific.

a. *Precision*: minimal vagueness, ambiguity and metaphoricity, so as to avoid misunderstandings and discourage fruitless debates about meanings.

b. *Communicability*: not exclusive to a brotherhood of initiates.
c. *Non-triviality*: not found in the body of ordinary knowledge, hence tolerant of some counter-intuitive original ideas.
d. *External consistency*: compatibility with the bulk of extant knowledge, hence ability to mesh in with other bits of knowledge.
e. *Testability in principle*: ability to confront empirical data, and thus capable of being either corroborated or falsified, however indirectly.

To clarify the above points, let us briefly examine the scientificity credentials of two widely held beliefs: psychoneural interactionism, and economic rationality.

Interactionism is "the theory that mental and physical states interact" (Popper & Eccles 1977: 37). The same doctrine also holds that, far from being on equal terms, "the [human] body is owned by the self, rather than the other way round" (op. cit., p. 120). Note the following flaws in the preceding quotes. First, the notions of *state* and of *interaction* are being wrongly used, for they are well defined only for concrete things, such as different brain sites, e.g. the prefrontal cortex and the motor center (see Bunge 1977). Second, the notion of *ownership* is legal, hence out of place in a scientific text except as a metaphor for proprioception or its temporary loss. Admittedly, dualism is falsifiable by any mental event occurring outside brains. But there is no scientific evidence for such events.

In short, the Popper–Eccles opinion on the question in hand is vague and consequenly unscientific. Worse, this view is two and a half millennia behind Alcmaeon's clear and fruitful hypothesis, that all mental occurrences are brain processes. Since this conjecture, which used to be called 'the identity theory,' is the philosophical flashlight of cognitive neurocience — the most advanced phase of psychology — the Popper–Eccles doctrine is not just unscientific. It is also outdated and a hindrance to the advancement of the sciences of the mind.

Our second example is the principle of economic rationality, which is common to all the rational choice theories proposed in the social studies over the past two centuries. This principle states that rational actors behave so as to maximize their expected utilities. (The expected utility of action a equals the product of the probability of a times the utility or benefit of a to the actor.) The probabilities and utilities in question are personal or subjective, so that they must be assigned arbitrarily, unlike objective rewards and punishments such as food pellets and electric shocks respectively. This feature renders expected utilities incorrigible, and the economic "rationality" principle at once imprecise and untestable.

Still, if we watch real people around us or in a lab while playing the "ultimatum" game, one finds that most of us share some of our winnings with fellow players, and even risk being punished for upbraiding individuals guilty of unfair or cruel actions. In short, most of the people we deal with in real life are reciprocal altruists rather than the psychopaths admired by the pop philosopher Ayn Rand and her star pupil Alan Greenspan, the top US banker during one decade (see, e.g., Gintis *et al.* 2005).

3.5 Excursus: From Wöhler's Serendipity to Ioannides's Bombshell

Let us quickly review two of the most unexpected and unsettling outcomes of modern science: Friedrich Wöhler's synthesis of urea, and John Ioannides's disappointing evaluation of biomedical research.

In attempting to synthesize ammonium cyanate in 1822, Wöhler obtained urea as a byproduct. This result was unexpected and unsettling, because urea was believed to be a manifestation of the vital force, a spiritual entity exclusive of living beings, according to the vitalist party ruling at the time. But Wöhler's intention was not to falsify this myth: this result was just a "collateral casualty" of his research.

Thus Wöhler's serendipitous finding falsified a widely held bimillennary belief: that, contrary to Hippocrates's hypothesis, "organic" and "inorganic" matter were radically different from one another. Moreover, the new science of biochemistry was born overnight, and modern pharmacology followed shortly therafter, while materialists rejoiced at the first serious threat to vitalism (e.g., Engels 1941).

Not given to philosophical speculation, Wöhler was not particularly impressed. But Berzelius, his mentor and friend and the greatest chemist of the day, was deeply shaken because he had been preaching vitalism all his life. It took him a while to admit the chemical and philosophical revolution that his beloved disciple had unleashed by accident (see Numbers & Kampourakis 2015: 59).

3.6 The Computer's Roles

No one denies the value of the computer in solving computational problems, in processing mountains of empirical data in hopes of uncovering trends, and in simulating real processes. For these reasons, it is justifiable to talk about the *computer revolution* brought about in the late 1970s by the spread of personal digital computers.

However, it is well known that many radical innovations have had harmful consequences along with their benefits, if only because they forced unexpected changes upon old habits. In particular, the computerization of scientific research has swelled the volume of small-caliber and even trivial findings, and the correlative promotion of competent and diligent craftsmen to the rank of scientific investigators.

Suffice it to recall the explosion of superficial and short-lived correlations in biomedical research and in social studies, particularly after software packages for correlation analysis arrived in the market ca. 1970. Using such devices, almost anyone can now study the linear correlation between any two variables picked at random. This is why the biomedical literature keeps swelling with the addition of weak and often short-lived papers purporting to prove that this substance or that activity is a "risk factor" for a given medical "condition."

Such is the kind of paper read by the normative social epidemiologists who design the sanitary policies adopted by public-health authorities. Thus our state of health is increasingly in the hands of biostatisticians who are not interested in the chemical, biological, or social mechanisms underneath the said correlations. Worse, many of these results turn out to be spurious: "most published research findings [in biomedical research] are false" (Ioannidis 2005). This result should not have surprised anyone given underlying empiricist assumption, that scientific research consists in data finding and processing — a byproduct of the computer cult.

In fact, Clifford Truesdell (1984) did warn us against the computer cult, which enjoins people to engage in the mindless search for empirical information and blind data processing without concern for the possible underlying mechanisms of action and the corresponding high-level laws. Thus, ironically, the preference for risk-free research projects is risky, in that it inhibits originality. The mechanism underlying this process is obvious: computers cannot generate new ideas, for they work only on algorithms, that is, rules for processing existing information. Consequently, the best a computer scientist can achieve is to invent better computer programs: he is a software engineer rather than a student of either nature or society.

A recent instance of risk-avoidance research is the massive computer-aided study of 6.5 million biomedical abstracts, in an effort to discover the dominant "research strategies" — the authors's name for types of research problem (Foster *et al.* 2015). The main finding of this study is that, as Kuhn (1977) said all along, "scientists' choice of research problems is indeed shaped by their strategic negotiation of an essential tension between productive tradition and risky innovation."

This result invites the following questions. First, how does the computer distinguish between highly original projects, such as the one that that gave us genetic editing, from a sure-thing one such as sequencing one more genome?

Second, why assume that the finished paper concerns the same problem that sparked the research, given the many unforeseeable connections and ramifications that may interpose between start and finish — unless of course the problem in hand is of the routine kind?

Third, how may one foresee whether the finding will help keep the boat afloat or scuttle it?

Fourth, what justifies ruling out both sheer curiosity and the supervisor's ability to choose an original and soluble problem, as well as to either steer good work or get rid of an unpromising student?

Fifth, how may one ascertain that a paper's main author contributed anything besides financial support or even his name?

A seasoned mentor is likely to hold that the actual process of problem choice is roughly as follows. Every research community contains scientific teams led by senior investigators who at any given time are working on a handful of projects. (For example, the 1,377 persons who won the Nobel Prize for their work on neutrinos were distributed among five different teams.) Anyone curious about one or more of these projects will approach its leader, and the two parties will discuss the possibility of the candidate joining his/her team. The decisive factors in this negotiation are the candidate's ability, dedication and determination, the leader's interest in recruiting him/her, and the available resources, from lab space to money.

Original doctoral theses take between two years and forever, with an average of six years. The question of the "essential tension" may not arise during such negotiations, for students usually learn from the grapevine who are the more productive, helpful, prestigious, and best-endowed mentors.

Still, there can be no guarantee that the chosen problem is indeed novel, much less that the proposed solution be interesting and likely to lead to further research. Just think of the old cynical comment on doctoral dissertations in humanistic studies: they are transfers of skeletons between cemeteries.

3.7 Demarcation Again

Popper's falsificationism has become increasingly popular in recent years, particularly at an unexpected place: the court of law where experts are expected to contribute scientific evidence. But an examination of any approved grant proposals submitted to a granting agency will hardly find projects whose main goal is to refute a belief. Refutations, when they occur, are "collateral damages," to use a military euphemism. For example, nobody set out to confute the thesis of the fixity of species: this negative result was just an unintended consequence of taking strange fossils seriously, namely as remains of organisms of extinct species, rather than as jokes of nature (*ludi naturae*). Likewise, epigenetics did not result from attempting to debunk genetic determinism, but from a chemical analysis of chromosomes of organisms subjected to unusual stresses.

To sum up, at any time "living" science boils down to the set of ongoing research projects. No research, no science. This is a sufficient reason for disqualifying esoteric and pseudoscientific beliefs and practices, as well as for putting on ice the opinions of celebrities about matters they have not researched.

The above conception of science also helps us answer questions of the "How-do-you-know?" form. Karl Popper dismissed them as unimportant because a person or a group may learn something from many different sources, from hearsay to textbook to a refereed individual paper to a whole spate of recent scientific papers. True, sometimes we learn falsities even from the most authoritative sources. But further research may correct scientific errors, whereas ideological howlers may persist for centuries if supported by the powers that be or by fanatic sects.

We seldom have the ability or the time to evaluate the credentials of all the knowledge claims we use for daily life purposes. But when building on such a claim in a research project, or when offering it as evidence in a court of law, we are expected to have subjected such a claim to a whole battery of truth tests. In short, when truth is of the essence we have the moral obligation of revealing how we learned that such-and-such piece of information is true to some degree.

Of course, not all research qualifies as scientific. For example, mere observation does not, because it does not involve the control of the relevant variables, as Claude Bernard argued in 1865. And only true, or at least likely hypotheses and theories, tell us *what* to observe, in particular which variables are worth measuring or wiggling. For example, physicians had no reason to take the pulse before Harvey showed that it is an indicator of heartbeat. Likewise, during centuries venereal diseases were regarded as skin ulcerations, hence as demanding the attention of dermatologists, until the disease's cause — infection with the spirochaete bacterium — was discovered in 1905. Once again we find that observation alone, without the help of sound hypothesis or theory, cannot advance scientific knowledge.

Although research projects are carried out by individuals or by groups, they should be impersonal, and therefore replicable. Only unique events, such as the emergence of radically new compounds, organisms, technologies, or social orders, are admissible exceptions to the replicability rule. In addition, this rule should be used sparingly in light of the dozens of alleged replications around the world following the original cold fusion fiasco in 1989. Hasty attempts to acquire or preserve reputation may end up in disrepute.

The research projects going on at any given time are constituents of a *system*, not just a set, because every new project builds on previous findings and, if successful and intriguing, it may suggest further projects in the same field or an adjoining one instead of ending up in a cul-de-sac.

The idea of Bacon, Hegel, and Husserl, of a science without presuppositions, is wrong, for we always take for granted a number of acquired ideas. Without some of them we could not even state radically new problems. And some of those received ideas are so firmly entrenched in our background knowledge that they are hardly examined, even though some of them may turn out to be false.

The preceding suggests the need for more studies in the foundations and philosophy of *scientia ferenda* or science in the making, instead of *scientia lata*, or done.

CHAPTER 4

SCIENCE AND SOCIETY

The classical historians of ideas have rightly been criticized for focusing on peaks rather than on mountain ranges, which is like confining a city to its skyscapers. By contrast, Marxists and the postmodernists have tended to exalt teamwork at the expense of individual talent, and even to claim that "society thinks through the individual," as if societies had collective brains with their full complement of memories and theories.

4.1 From Lonely Genius to Research Team

During most of the 20th century, Albert Einstein was generally regarded as the greatest scientist of the century, or even of history. In recent times, there have been mean attempts to cutting him down to size, to the point of claiming that his relativity theories were the product of the collective effort of his circle of intimate friends, in particular his first wife Mileva Maric, his former fellow student Marcel Grossman, and his friends Conrad Habicht, Maurice Solovine, and Michele Besso, his only colleague at the Swiss Patent Office. Others have noted that Hendrik Lorentz and Henri Poincaré had known the Lorentz transformations, which look like the signature of *SR*. Is there any truth in all of this?

Einstein himself gave the right answer, which can be compressed into the following sentences. First, the special theory of relativity (*SR*) was the culmination of Maxwell's classical electrodynamics. No wonder then that others, particularly Lorentz and Poincaré, came close to it. But they lacked Einstein's youthful courage to rebuild mechanics so that its basic laws would be Lorentz-invariant, like those of electrodynamics.

As Einstein himself said, *SR* could have been built by several others; by contrast, only he could have built general relativity (*GR*), his theory of gravitation, if only because no one else was working on gravitation at that time.

Second, of course Einstein discussed his new ideas with his wife Mileva, a fellow if failed physicist, as well as with his closest friends. But the latter's contributions were not equivalent: while Grossmann's was central, Besso's was peripheral. Indeed, Grossmann taught his friend the mathematical tool he needed to build *GR*, namely the absolute differential geometry, or tensor calculus; the scientific association of both men was so close, that they wrote a joint paper.

By contrast, Besso's role was, in Einstein's own words, that of a sounding board, or kibitzer, as one would say today. Moreover, Besso's attempts to convert his younger friend to Mach's phenomenalism and operationism were futile: while admiring Mach's experimental skills, as well as his sketchy relational view of space and time, Einstein, like Boltzmann and Planck, was a sharp critic of Mach's subjectivism and an early defender of scientific realism (e.g., Einstein 1950).

Some academic feminists have claimed that Einstein stole his *SR* from Mileva, his first wife. This claim is unsubstantiated, and it does not explain the fact that *SR* was only one of the four original ideas that he crafted in his *annus mirabilis* of 1905. It does not even explain why Mileva was not invited to join the informal Olympia Akademie that Einstein founded in 1902 along with Conrad Habicht and Maurice Solovine to discuss problems in physics and philosophy.

The academic feminists have also claimed that Hypatia, the neo-Platonist murdered by a Christian mob, had been a great mathematician, but they do not tell us what she had accomplished. More recently, the same group has claimed that the British crystallographer Rosalind Franklin deserved sharing the Nobel Prize with Francis Crick and James Watson for discovering the structure of the DNA molecule. While there no doubt that Franklin did contribute to that discovery, it is also true that others, in particular Linus Pauling, made even more important contributions to the same project, but only Crick and Watson came up with the prized answer.

In sum, productive scientists do not work in isolation but as members of networks rooted in the past. Not even the reclusive Newton was an isolated genius. In fact, we know from Samuel Pepys's diaries that his ideas, mainly his problems, were discussed in Pepys's circle. But Newton's colleagues did not share his interests, let alone his grand vision, and his Unitarianism prevented him from having students. In short, Voltaire was quite right in worshipping Newton.

4.2 The Research Team

Up until recently, most research projects involved a single investigator assisted by a few collaborators whose contributions were acknowledged at the end of the report. From about 1950, the typical research project has involved the principal investigator together with a few collaborators, usually his/her doctoral and postdoctoral students, all of whom were recognized as coworkers and given equal credit.

Research teams in experimental particle physics, astrophysics, genetics, and biomedical research grew to involve a hundred or more investigators. Occasionally a scientific paper would be signed by a hundred or even a thousand researchers, so that the list of their names, ordered alphabetically, would take up a whole page of a scientific journal. This kind of collaborative research was dubbed Big Science (de Solla Price 1963), to mark its differences with the previously dominant Little Science, where the principal investigator took all the credit and assumed all the responsibilities, starting with the grant proposal.

But theoretical research, however important, kept being essentially an individual's task done in solitude although discussed in seminars. The few times a theorist was appointed to lead a big research project, he ceased to do original work, and was unable to resume research when his managerial task ended. J. Robert Oppenheimer's life before and after his stint as the scientific director of the Manhattan Project (1942–45) is a case in point.

4.3 Scientific Controversy

Unlike religious scriptures and ideological programs, scientific research projects are open to criticism from beginning to end. But, unlike philosophical and literary criticism, scientific criticism is sought by the researcher, for it is characteristically constructive: it is performed by colleagues who share a background and aim at perfecting the work under consideration rather than killing it at birth with biting remarks.

Isaac Asimov called *endoheresy* this kind of criticism, in contrast to the *exoheresies* typical of the enemies of the scientific stance. Clear cases of endoheresy were Maxwell's criticism of Ampère's action-at-a-distance electrodynamics, Einstein's criticism of classical mechanics, and Stephen J. Gould's criticism of the "nature makes no jumps" dogma. In all of these cases, criticism paved the way for deeper and more comprehensive theories, as well as for novel experiments.

A clear case of exoheresy, or destructive criticism by non-professionals, is the campaign against the French Enlightenment waged by the Frankfurt school and by such popular writers as Michel Foucault and Bruno Latour, the standard-bearers of the massive attack on Robert Merton's (1973) pro-science sociology of science.

Another famous exoheresy was the attack on "bourgeois science" by Marxist philosophers during the first half of the 20th century. This criticism originated in misunderstandings of the scientific novelties of the day, and it stunted the development of science in the so-called Socialist camp. The worst aspect of such destructive criticism was its vindication of the obscurities and absurdities of Hegel's dialectics, such as the definition of becoming as the "dialectical synthesis" of being and nonbeing. This ecoheresy had a boomerang effect: it discredited the whole of Marxism, instead of rejecting only the bad philosophy in it.

However, historical materialism, the Marxist view on social change, may be viewed independently from dialectics, and as a materialist conception of history, that is, as the hypothesis that material interests, rather than ideas, are the prime movers of social action.

This conception has had a beneficial effect on the problems of the origin of life and hominization, as well as on anthropology, archaeology, and historiography (Barraclough 1979; Fontana 2011; Harris 1968; Trigger 2003). For example, historical materialism has suggested to anthropologists that they should start by finding out how their subjects make a living, instead of learning what they believe and how they entertain themselves. And it has sought the sources of domestic and international conflicts in material interests, such as the control over trade routes in antiquity, land in the Middle Ages, and oil in recent years.

For instance, historical materialists are likely to assume that the Thirty Years' War (1618–48) was not over religion, as is we were told at school, but over land, as shown by the fact that most of the soldiers under the Catholic Emperor Charles V were Lutheran mercenaries who took double pleasure in sacking the Pope's seat.

Another example of the salutary influence of materialism on historiography is the debunking of the myth that the secret services won World War II. The truth is that nearly every gain in the military intelligence of one of the sides was offset by a triumph of the other side. As the military historian Max Hastings (2015) has shown in detail, intelligence and disinformation work only when assisting the armed forces.

While intelligence did help win some battles, the war was won by the Soviet soldiers in Stalingrad, not by the Bletchley Park code-breakers. And the Japanese asked for peace terms after their civilians were bombed and sprayed with napalm, even before Hiroshima and Nagasaki were wiped out by nuclear bombs (Blackett 1949). Modern war uses as much brain as brawn, but it is not a spiritual pursuit. Consequently, neither waging nor understanding it is a hermeneutic exercise.

In short, historical materialism has been good for historiography, though not as good as the *histoire totale* of the *Annales* school led by Marc Bloch, Lucien Febvre, and above all Fernand Braudel (see Schöttler, 2015). These scientists started by studying the material resources, but did not neglect the political or cultural aspects. And they knew about class conflicts but did not share the Marxist thesis, that class struggle is the engine of history in all cases, not only in those of the Peasant Wars in Martin Luther's time, the French Revolution (1789–99), the Spanish Civil War (1936–39), and the Chinese Civil War (1927–49). Conflicts occur in all social systems, but systems result from cooperation.

4.4 Postmodernist Travesties

Up until the 1950s, the study of scientific communities had been a task of philosophers, sociologists, and historians of science intent on finding truths about science, that much-celebrated yet still elusive beast. Suffice it to recall the philosophical and historical studies by John Herschel, William Whewell, William Stanley Jevons, Karl Pearson, Henri Poincaré, Émile

Meyerson, Federigo Enriques, Pierre Duhem, Albert Einstein, the members of the Vienna Circle, Karl Popper, Morris Raphael Cohen, Eduard Dijksterhuis, I. Bernard Cohen, Joseph Needham, Charles Gillispie, Ernest Nagel, Richard Braithwaite, Eino Kaila, Aldo Mieli, George Sarton, and Robert K. Merton.

In his classic 1938 paper on "Science and the social order," published in the young journal *Philosophy of Science*, Merton had argued that the peculiarities of basic science are *disinterestedness, universality, epistemic communism, and organized skepticism* — not the doubt of the isolated researcher but the constructive scrutiny by a whole community.

Unlike his critics, Merton was not a dilettante but the first professional sociologist of science. His teachers had been the leading sociologists of his day — Pitirim Sorokin, George Sarton, and Talcott Parsons — as well as the amazing chemist, biologist and sociologist Lawrence Henderson, who had rescued and popularized the concept of a social system. Besides, partly thanks to his wife and colleague, Harriet Zuckermann, Merton got to know personally many Nobel laureates, who told him what made them tick, from whom they had learned, and how their respective scientific communities had now stimulated, now inhibited them.

In sum, around 1950 Merton was recognized as the most learned member of the science-studies communities. His studies were also the most balanced of them: Merton was the only one who, though not an idealist, stressed the disinterestedness of basic researchers; while not a positivist, Merton admitted the cumulative nature of science, and, though not a Marxist, he stressed the social embeddedness of the scientific community, as well as the political pressures it was subjected to.

Suddenly, in 1962, in his best-seller *The Structure of Scientific Revolutions* Thomas S. Kuhn, an obscure scientist, held that scientists do not seek truth because there is no such thing, nor is there a body of knowledge that grows and is being repaired and made ever deeper. His central thesis was that, once in a while, there occur scientific revolutions that sweep away everything preceding them. Moreover, such radical changes would not solve long-standing scientific problems, but would respond to alterations in the *Zeitgeist* or cultural fashion of the day. Hence, scientists would neither confirm nor confute anything: as his friend and comrade in arms Paul Feyerabend declared, "anything goes." In short, these nihilists challenged the prevailing view of science. Henceforth any amateur with enough *chutzpah* qualified for a job in one of the many "science-studies" centers or "science and society" programs that have proliferated over the past few decades.

This counter-revolution was so massive and so sudden, that it took the academic community by storm and by surprise (see Bunge 2016a). Since then, the so-called *science wars* have been waged, with more noise than light. The Australian David Stove (1982) was one of the very few philosophers to ridicule it, but the alternative he offered, a return to old-fashioned empiricism, did not persuade anyone. Only the publication of Alan Sokal's hilarious hoax "Transgressing the boundaries: Towards a transformative hermeneutics of quantum gravity," in *Social Text* , which had hailed the Kuhn-Feyerabend coup, told the public that they had been fooled by a troupe of clowns (Sokal & Bricmont 1997).

My own detailed studies on the philosophical roots of the Kuhn–Feyerabend counter-revolution (Bunge 1991; 1992; 1997) were hardly noticed by the philosophical community. Merton's realistic image of basic science has all but been jettisoned by most metascience students, both on the right and on the left, who reject the very idea of pure science.

Scientism, vigorous and prestigious one century ago, is now weak and discredited in the humanist camp, where Friedrich Hayek's biased definition of it as "the attempt to ape the natural sciences in the social ones" has prevailed. Another misconception of science popularized in recent times is Michel Foucault's grotesque characterization of science as "politics by other means" — a myth exploded by Dominique Raynaud's (2015) careful studies of a number of famous scientific controversies. He has shown that, even in the cases where political or religious power meddled, the controversies in question were about truth, not power. And in the end truth won out.

The main reason for this is that scientific research seeks original truth, not practical benefit — a goal of technology. For example, the debate on whether the quantum theory refers to physical objects or to measurement operations is purely cognitive: neither party in this nearly centennial debate has anything ideological at stake. By contrast, some of the controversies in social science raise ideological issues. For example, standard economic theory has rightly been blamed for the economic crisis starting in 2018, for ignoring disequilibria and praising selfishness; the idealist philosophy of the social championed by Wilhelm Dilthey is guily of ignoring the material needs and interests, particularly those of the poor; and Marxism has been accused of overlooking the individual's talent, in stating Marx's holistic dogma that "society feels and thinks through the individual."

That truth, not power, was at stake in all of these cases, as well as in those discussed by Raynaud, is a point in favor of scientism — Condorcet's thesis that whatever can be investigated is best studied using the scientific method. By the same token, it is also a point against the intuitionism inherent in the "humanistic" school, as well as against the sociologism typical of both the early Marxist and Durkhemian schools in the sociology of knowledge, which all but ignored talent.

Productive social scientists do not regard their discipline as a *Geisteswissenschaft* (spiritual science) requiring a method of its own, such as the *Verstehen* or empathic understanding exalted by Wilhelm Dilthey in his 1883 anti-scientistic manifesto. To be sure, putting oneself in A's position (no mean feat!) may help explain why A thought or did B, but it does not explain B itself. Likewise, placing A in his/her social context may help explain why B was either recognized or suppressed by the establishment but, again, it does not account for B itself. Axiomatics, which identifies the central ideas of a theory, helps understand why science is self-propelling and self-serving. More on this anon.

CHAPTER 5

AXIOMATICS

Usually we think in a logically disorganized manner: we do not distinguish basic from defined concepts, assumptions from consequences, or constitutive from regulative principles. When proceeding in this spontaneous fashion we can advance quickly, but we may inadvertently introduce or conceal controversial or even false assumptions, that imperil the whole construction. Axiomatics is intended to avert such catastrophes or repair buildings erected in a hurry.

5.1 Intuitive and Axiomatic Reasonings

To axiomatize consists in subjecting a theory originally built in an intuitive or heuristic fashion to the following operations:

a. To *exactify* intuitive constructs, that is, to replace them with precise ideas, as when substituting "set" for "collection," "function" for "dependence", and "derivative with respect to time" for "rate of change;"
b. To *ground* and, in particular, to *justify* the postulates and bring hidden assumptions to light — assumptions that, though seemingly self-evident, may prove to be problematic; and
c. To *order deductively* a bunch of statements about a given subject.

These three tasks are interdependent: when exactifying an idea one may discover that it implies others or is implied by others, and when ordering a handful of propositions, one may discover a missing link or an unjustified premise or consequence. For instance, since Kant, the empiricists and the phenomenologists equated existence with possible experience, they tacitly assumed that humans have always existed, while actually our species arose only about two million years ago.

The main interest of axiomatics for philosophers is that some of the tacit assumptions of the intuitive formulations of important theories may be philosophical rather than either logical or factual. For example, set theory has been axiomatized with as well as without the axiom of choice, which constructivists reject because they demand a precise constructive choice function instead of the promise of one, as implied by the phrase "there exists a function such that …"

Second example: in 1925 Heisenberg published his matrix quantum theory, claiming that it included only measurable magnitudes. But in 1941 he admitted that this was not true, and proposed his S-matrix theory, which in fact was closer to experiment but, alas, it was also incapable of solving any problems without the help of standard quantum mechanics, as a consequence of which it was soon forgotten — even by its author, who does not mention it in his 1969 memoir. A realist philosopher might have spared him those disappointments, but the only philosopher in his Leipzig seminar was a Kantian who supported the Copenhagen "spirit," the idea that there are only appearances (to someone).

Ordinarily one axiomatizes with one of these goals: to unify previously disconnected findings (Euclid's case), to deepen the foundation of a research field (Hilbert's case), or to eliminate paradoxes. For example, in 1909 Ernst Zermelo axiomatized set theory to avoid the paradoxes inherent in the naïve theory built by Bolzano and Cantor, and that imperfection had kept Frege and Russell awake while confirming Poincaré's skepticism. Another example: the mathematician Constantin Carathéodory wished to gather, cleanse, and order logically the scattered thermodynamic findings of Carnot, Clausius, and Kelvin. By contrast, my own motivation for axiomatizing the two relativities and quantum mechanics (Bunge 1967a) was to rid them of the subjectivistic elements that had been smuggled into them by the logical positivists.

Regrettably, the price paid in the first two cases was staggering. Indeed, Zermelo's axiomatics deals with sets of sets, so that it gives preference to the notion of species to that of individual. This Platonic bias renders it useless in the factual sciences, which went on using, if at all, naïve set theory (e.g., Halmos 1960), for it starts with individuals.

As for Carathéodory's axiomatics, it was restricted to reversible and adiabatic processes, which are hard to find in either nature or industry, where irreversible processes, such as those of dilution, diffusion, explosion, implosion, and heat transfer prevail. Thus Carathéodory achieved mathematical rigor by taking *dynamics* out of *thermodynamics*.

That is why the subsequent contributions to the field, such as those of Lars Onsager, Ilya Prigogine, and Clifford Truesdell, owed nothing to Carathéodory's thermostatics. The latter remained a tool for taming engineering students, giving a few philosophers of science the occasion to show that they knew neither philosophy nor science.

Many of the physics students of my generation learned classical thermodynamics in Fermi's textbook, and statistical mechanics in Landau and Lifshitz's. The latter taught us Boltzmann's lasting lesson: that, far from being a fundamental and isolated discipline, thermodynamics was the culmination of statistical physics, which explained heat as a macrophysical effect of the random motion of lower-level entities. In addition, it reminded us that the most interesting macrophysical processes, those of self-organization, occur in open systems, where neither of the first two famous laws is satisfied.

An unforeseen consequence of Carathéodory's axiomatics was that some pedagogues misinterpreted it as a theory of *states in themselves* rather than states of a thermodynamic system, such as a heat-transfer device (e.g., Falk & Jung 1959; Moulines 1975). This plain mistake suggested another: that natural science is not about material things (Moulines 1977). But of course even a mathematician such as Carathéodory, when writing *Zustand* (state), presupposes that this is the state of the *concrete* system in question, since it makes no sense to speak of the state of abstract objects. See the alternative axiomatization proposed by Puccini *et al.* (2008), who know what thermostatics is about, namely closed macrophysical systems, where no qualitative novelties occur.

5.2 The Models Muddle

All model theorists, such as Alfred Tarski, know that their models are examples or interpretations of abstract theories (or formal systems) such as those of graphs, lattices, and groups — hence unrelated to the theoretical models devised by scientists and technologists, which are special theories, such as that of the simple pendulum. Thus the entire model-theoretic (or structuralist) approach to theoretical physics, adopted by Joseph Sneed (1971) and his followers, such as Moulines and Stegmüller, is the fruit of an equivocation, as would be regarding ring theory as dealing with wedding bands, onion rings, and the like.

However, let us go back to our central theme. Perhaps the greatest virtue of axiomatics is not that it enhances formal rigor but that it uncovers the tacit assumptions in the intuitive formulations, such as that the laws of thermodynamics hold regardless of the number of basic components, which Boltzmann had doubted when he allowed for violations of the second law in the case of small numbers. Another, philosophically more interesting case, is the claim that things acquire their properties only when observed, which inadvertently assumes that the universe was not born until the first modern laboratory was set up.

A better-known example is this. The operationalist demand that all the physical concepts be defined in terms of laboratory operations implies distinguishing two different masses of a body: the inertial, which occurs in Newton's law of motion, and the gravitational, inherent in his law of gravitation. But any correct axiomatization of classical mechanics will contain a single mass concept, to allow for the cancellation of m in equations such as that for a falling body: "$mg = GmM/r^2$."

This does not preclude adding the remark that mass has three *aspects*: as a measure of inertia, of gravitational pull, and of quantity of particulate matter. Likewise the electrodynamic potential A has two faces: it generates the field and it accelerates electrically charged matter. And when writing de Broglie's formula $\lambda = h/p$, one evokes both the corpuscular (p) and the wavelike (λ) metaphors, but one does not claim that there are two kinds of linear momentum.

Even Einstein, otherwise an outspoken realist, fell into the operationist trap when he admitted Eötvös claim that he had measured both the inertial and the gravitational masses of a body, finding that they are the same, while actually he had measured a single property, mass, with two different methods. Likewise, one may measure a time lapse with a clepsydra, a pendulum, a spring clock, or other means, which does not prove that there are multiple times. Distance, temperature, energy, and most other magnitudes are parallel.

The ultimate reason for the one-to-many correspondence between magnitudes and their measurements is that the properties of real things come in bundles, not in isolation — a metaphysical principle. For the same reason, and also because every measurement apparatus calls for its own special theory and its own indicator, it would be foolish to tether any general theory to a particular measurement procedure.

Something similar holds for the social theories and techniques. For example, even a partial axiomatization of standard economics suffices to discover its least realistic assumptions: that markets are free and their members rational (see Bunge 2009a). Another case in point is this: if handled intuitively, one runs the risk of treating the key features of a social group one by one rather than in combination with other properties of it. For example, those who claim that liberty beats all the other social values ignore that there can be no liberty where power is held by privileged individuals, be they tyrants, tycoons, or priests.

The systemic or integral approach to any fragment of reality suggests favoring theories that emphasize the key variables and the connections among them. For instance, liberty will be linked with equality and solidarity, as the French revolutionaries claimed in 1789, and it may be added that this famous triad rests on another, namely occupation, health, and education. It is up to political theorists to imagine ways of constructing such a hexagon, and to the more rigorous among them to construct theories clarifying and inter-relating the six variables in question (see Bunge 2009c).

In sum, intuitive or heuristic thinking can be creative and fast, but it may be marred by muddled concepts or false assumptions or presuppositions, which in turn are bound to entail wrong conceptual or practical consequences. In cleaning and ordering premises and arguments, axiomatization may save us from such mistakes and the corresponding barren controversies.

5.3 Axiomatic *vs.* Heuristic Formulations of Theories

The terms 'hypothesis' and 'theory' are synonymous in ordinary language. Not so in the exact sciences, where a theory proper is a hypothetic-deductive *system* of propositions closed under deduction, and whose components support one another, so that the outcome of the empirical test of any of them affects the standing of the others. Ideally, all the knowledge about any domain is contained in one or more theories plus a set of empirical data.

For the most part, scientific theories are formulated in a heuristic manner, so that anyone feels entitled to add any opinions or even to interpret them as they wish. For example, some authors say that quantum physics is about microphysical entities, whereas others claim that it admits no levels; and, whereas some restrict its domain of validity to objects under observation, or even to object-apparatus-observer wholes, others admit that it also holds outside labs — for instance, in stars. Again, some authors deal with the Heisenberg *principle* on one page, only to *prove* it on another. And most authors freely exchange 'indetermination' and 'uncertainty,' so that the reader is unsure whether Heisenberg's inequalities constitute a law of nature or just an opinion about the limitation of human understanding. Thus one may be left with the impression that the author did not know what he wrote about.

Only a suitable axiomatic formulation of quantum mechanics may prove that Heisenberg's inequalities constitute a theorem, not a principle. And only a realist axiomatics can claim that, if true, those formulas will constitute a law of nature, not a limitation on our knowledge of it. It will accomplish all this by stating from the start that the theory is about real existents, not about observations by means of the mythical Heisenberg microscope, or the no less mythical clock-in-a-box that Bohr imagined to "derive" his time–energy inequality, which is no part of the theory (Bunge 1970).

In proceeding in an orderly manner, the axiomatizer will be helped by the old logical finding that no set of empirical data, however bulky, can prove a general statement — if only because the theory contains predicates that do not occur in the empirical data. Much the same holds for the mostly groundless opinions on scattered in the literature. It took me two decades to realize that only reasonings from principles could justify any assertion of the kind. This is why I undertook to axiomatize several contemporary physical theories and free them from unjustified philosophical assumptions (Bunge 1967a).

Other physicists have updated or expanded that effort (Covarrubias 1993; Pérez-Begliaffa *et al.* 1993; 1997; Puccini *et al.* 2008). Physical axiomatics was thus a fruit of the union of philosophical realism with the drive to replace quotation with argument, and a bunch of stray statements with a single axiomatic system.

5.4 Dual Axiomatics: Formal and Semantic

Euclid (ca. 300 C.E.) is likely to have been the earliest axiomatizer: he collected and ordered all the bits of geometric knowledge accumulated in the course of the preceding centuries. Two millennia later another giant, Bento Spinoza, revived axiomatics for philosophical purposes. And around 1900 David Hilbert, Giuseppe Peano, Alessandro Padoa, and Alfred Tarski updated and applied the Euclidean format. The latter may be summarized as follows:

Primitive or undefined concepts.

Postulates or axioms.

Lemmas, or statements borrowed from other fields.

Theorems.

Corollaries.

In some cases the definitions are given right after listing the primitives, whereas in others they are introduced further down, as new concepts occur in theorems. Occasionally the foundation of a theory is condensed into a single axiomatic definition, as will be seen in Section 5.8.

All of the above is fairly standard fare and of no great mathematical interest, since anyone acquainted with a given theory can axiomatize it without tears, as long as s/he does not question the philosophical motivations underlying the preference for one choice of primitives over another. Being purely structural, the mathematical formalism of an axiomatic system calls for no extramathematical elucidation.

Consider, for instance, the mathematical formalism of Pauli's theory of spin one-half particles, such as electrons. The core of this formalism is the spin vector $\sigma = u_1\sigma_1 + u_2\sigma_2 + u_3\sigma_3$, where the u_i, for $i = 1,2,3$, are the components of an arbitrary unit vector, while the corresponding σ_i are the 2×2 Pauli matrices, which are implicitly defined by equations such as $\sigma_1\sigma_2 - \sigma_2\sigma_1 = 2i\sigma_3$. So far, this is only a piece of undergraduate algebra.

But trouble is bound to start if someone asks how to interpret σ in physical terms, that is, if s/he asks what is the physical property called 'spin?'

The vulgar answer is that is "the intrinsic angular momentum," all the more so since the equations defining it are similar to those satisfied by the quantum counterparts of the orbital angular momentum $L = r \times p$. But a moment's reflection suffices to realize that this answer is an oxymoron, for most of the said particles are also assumed to be pointlike, and points cannot revolve around themselves. In short, electrons and their ilk do not spin any more than they weave.

A more cautious if elusive answer is that spin is "responsible" for the Zeeman multiplication of the spectral lines of an atom when embedded in a magnetic field, as well as for the splitting of an electron beam when entering the magnetic field in a Stern–Gerlach apparatus. But these answers do not tell us anything about the underlying mechanism or modus operandi.

In my view σ is a very useful *mathematical auxiliary* with no physical meaning. What does have such meaning is the *magnetic moment* $\mu = \mu_B \sigma$, where $\mu_B = eh/4\pi mc$ is the Bohr magneton. In metaphorical terms, μ_B is the physical flesh attached to the mathematical bone σ. Moreover, that flesh is magnetic and unrelated to spinning, which is a dynamical process. Thus, 'spin' is a misnomer: electrons and their ilk are not like spinning tops but like magnets. And the theories in factual science resemble cutlets, in that they are mathematical skeletons with chunks of factual meat attached to them.

This account coheres with Heisenberg's explanation of the difference between ferromagnetism and paramagnetism in terms of the alignment, total and partial respectively, of the magnetic moments of the valence electrons of the atoms constituting the material in question.

Likewise, what accounts for the splitting of an electron beam entering the magnetic field of intensity H in a Stern–Gerlach apparatus is not σ but μ: the electrons with a magnetic moment parallel to the field ("spin up") acquire the additional energy $\mu_B H$, whereas the antiparallel ones ("spin down") transfer the same amount of energy to the external field.

In short, the physical property or magnitude in question is not the spin, a non-dimensional mathematical bone, but the elementary magnetic moment, which is affected by an external magnetic field and explains, among other things, the multiplication of the spectral lines of an atom immersed in a magnetic field. Thus the Zeeman effect is explained in terms of the perturbation that H causes on the electrons' intrinsic magnetism, not on their imaginary spinning.

Let us now return to the original subject, namely the *pairing of every mathematical axiom of a factual theory with a semantic assumption assigning it a factual meaning* (that is, reference and sense). Such an assumption is necessary to learn what is being assumed about what. Ordinarily the context will suffice to perform this task. But some cases, such as those of the terms 'energy,' 'mass,' 'entropy,' 'potential,' 'spin,' 'state function,' and 'information,' are far from trivial, and have originated controversies lasting decades. The reason is of course that pure mathematics is not about real things, even though some mathematical concepts, such as those of derivative and integral, were born with geometrical and dynamical contents. Only the addition of a semantic assumption may disambiguate or "flesh out" a mathematical formula occurring in a factual discourse.

The formalist school started by the McKinsey *et al.* (1953) paper on the axiomatization of classical particle mechanics overlooks semantics. In identifying a living body with its skeleton, the formalists fail to explain why the same mathematical concepts occur in many diffferent research fields, though mostly with different meanings. This is why they have not participated constructively in the controversies provoked by the two relativities, quantum mechanics, genetics, psychology, or economics.

Physicists have not objected to the mathematical formalism of quantum mechanics: the spirited debates about it over nearly one century have concerned its interpretations. This question is so important, that Max Born earned the Nobel Prize basically for proposing the so-called statistical (actually probabilistic) interpretation of the famous ψ.

However, let us be fair: McKinsey's foundational paper of the Suppes–Sneed–Stegmüller structuralist school was just ill-timed: had it appeared two and a half centuries earlier, it might have shed some light on the discussions among Newtonians, Cartesians, and Leibnizians, that Newton's *Principia* had provoked in 1687. And even had it been published as late as in 1893, it would have saved college physics teachers from Mach's wrong definition of "mass," since the said article uses Padoa's method to prove the independence of this concept from the remaining primitives of Newtonian particle mechanics.

We shall call *dual axiomatics* the one that accompanies every key mathematical concept with a semantic hypothesis specifying its reference and sketching its sense (Bunge 1967d; 1967e). We call it *hypothesis* and not *convention* or *rule* because it can be overthrown by observation or experiment. For example, Hideki Yukawa's pioneer meson theory of 1935 was initially assumed to concern mu mesons, until it was found to describe pi mesons.

When the semantic component is overlooked, one runs the risk of incurring mistakes like the so-called Aharonov–Bohm effect. This consists in believing that the electrodynamic potential A related to the magnetic feld intensity by $H = \nabla \times A$ is just a mathematical auxiliary, because it may happen that $A \neq 0$ but $H = \nabla \times A = 0$. An operationist will hold that such an A has no physical meaning because it does not affect a magnetized needle, but a realist may remind her that A will decelerate an electron but accelerate a proton, by altering the particle momentum in the amount $-(e/c)A$. For this reason, a realist is likely to recommend axiomatizing classical electrodynamics by starting with the four current densities and the corresponding potentials rather than the field intensities, even though they represent aspects of one and the same thing — the electromagnetic field (Bunge 2015).

In sum, we reiterate the axiomatization strategy proposed in earlier publications (Bunge 1967a; 1967c; 1967f), which differs from the structuralist one defended by Suppes, Sneed, Stegmüller, Moulines, and other philosophers used to dining on bare ribs.

This formalist stance, which ignores the semantic side of scientific theories, and skirts the controversies this side generates, was the target of my criticisms of one of Stegmüller's books (Bunge 1976). But the oldest and most widely read journal in philosophy of science rejected my laudatory review of Truesdell's (1984) massive demolition of that school. Speak of party philosophy on the other side of the Iron Curtain!

5.5 The Vulgar View on Quantum Physics

Since its inception in 1900, quantum physics has been the object of many spirited controversies. What was being questioned was its physical interpretation. But, because the theory was axiomatized only in 1967, most of those controversies were about just a few components of it, and they were basically clashes of opinions among a few leaders of the physics community, mainly Bohr and his many followers on one side and a handful of dissenters led by Einstein on the other.

Worse, most of the discussants confused philosophical realism — the thesis of the reality of the external world — with what I call *classicism* (Bunge 1979a). This is the opinion that the quantum theory is seriously flawed because it does not calculate the precise positions and trajectories of quantum objects (Einstein *et al.* 1935). These critics did not admit that the ultimate constituents of the universe might not have such properties. Scientific realists, by contrast, do not commit themselves to any particular physical hypotheses: they just assert that physicists are expected to study things in themselves.

Contrary to scientific realism, the "official" opinion, held by Bohr's Copenhagen school, holds that electrons and the like come into being only upon being observed or measured. The most extreme among the orthodox hold that "[t]he universe is entirely mental" (Henry 2005) or even sacred (Omnès 1999). Such strange ex cathedra revelations did not result from an examination of the principles of quantum mechanics. When these are identified and examined, a sober picture emerges: quanta are certainly quaint from a classical-physics viewpoint, which is why they deserve a name of their own, *quantons* (Bunge 1967e). But they do not imperil the project of the Greek atomists, of uncovering the real and law-abiding universe beneath appearances and beyond myth.

We shall confine our discussion to the main issues that confronted physicists and philosophers before Schrödinger's cat stole the show, that is, during the formative decade between Heisenberg's foundational paper of 1925, and the Bohr–Einstein debate in 1935. It is pointless to rush to later issues before settling the earlier ones, which are still festering and, more to the point, are still being discussed *more theologico* rather than *more geometrico* (see Bunge 1956).

Most of the controversies about quanta revolve around three of its key concepts: the state (or wave) function ψ; the eigenvalues a_k of the operator A_{op} representing a dynamical variable ("observable" A, which occur in the equations of the form $A_{op}u_k = a_k u_k$; and the symbol Δ in the inequalities of the form $\Delta p \, \Delta q \geq h/2\pi$.

The subjectivist and objectivist (realist) interpretations in question are summarized in the following table:

Symbol	Copenhagen	Realist
ω	Measured object	Object in itself (quanton)
x	ω's position coordinate	Arbitrary point of space
$H(x,p,t)$	Hamiltonian	ω's energy
$\lvert\psi(\omega,x,t)\rvert^2\Delta v$	Probability of *finding* ω inside Δv when measuring ω's position at time t.	Probability of ω's *presence* inside Δv at time t.
a_k	Measured value of A with probability $\lvert u_k\rvert^2$	*Objective* value of A with probability $\lvert u_k\rvert^2$
ΔM	Uncertainty about M's value	Variance or mean square deviation of the really possible values of M

Every one of the above statements is a *semantic assumption*, and moreover a controversial one. Consider, in particular, the orthodox dispute over the very existence of the referent ω, or at least its existence separate from both observer and apparatus; realists adopt the view that probabilities represent possibilities of futures rather than either frequencies of past events or degrees of belief; and if the innocent-looking x is interpreted as the referent's position, it follows that its time derivative is the particle velocity, which in Dirac's and Kemmer's relativistic theories turns out to be the matrix $c\alpha$, whose eigenvalues are $+c$ and $-c$, which is absurd and consequently calls for a different coordinate (Bunge 2003b).

The realist interpretation does not state that the object ω of study is being observed all the time. Only an examination of the energy operator H can tell us whether the object under study is free or under the action of external actions, in particular the actions exerted by a measuring apparatus. In the latter case the said operator will include a term H_{int} depending on both ω and dynamical variables characterizing the apparatus. If the latter do not occur in H, as is the case when calculating energy levels of free atoms or molecules, talk of experimental perturbations is philosophical smuggle. (Wolfgang Pauli, though orthodox, admitted that some experimental devices, such as the spectrometer, are non-invasive.)

Show me your hamiltonian, and I'll tell you what it is about.

Note also that the orthodox view fails to specify the measuring instrument and the indicator the experimenter uses, as though universal instruments and indicators could be built. It is ironical that an approach said to follow closely laboratory procedures, actually involves a magician's top hat or, to put it more politely, "a 'black-box' process that has little if any relation to the workings of actual physical measurements" (Schlosshauer 2007: 334).

Any responsible talk of observables and actual measurement involves a detailed description of specific laboratory procedures. Such account, being tied to a specific setup and a specific indicator, has no place in a general theory such as quantum mechanics, just as statesmen have no right to stipulate the value of π, as the writers of the Indiana constitution did when they legislated that $\pi = 3.14$. Therefore it is wrong to claim, as Dirac (1958) did with respect to the eigenvalues of "observables," that they are the values that any experimental procedure must yield. If this were true, governments could close down all the laboratories, as René Thom proposed when he launched the biological interpretation of his "catastrophe" (singularity) theory.

5.6 Reasoning from Principles Instead of Quoting Scripture

Let us learn from Anselm of Canterbury, and argue from principles instead of walking on Scriptural crutches. In the case in hand, I submit that the only rational way to approach foundational dilemmas like the subjectivism/realism one is to examine the foundations of the theory in question, which in turn requires axiomatizing it. The rest is leaning on quotations, hand-waving, or preaching.

For example, the Heisenberg-like inequalities for an arbitrary pair of conjugate canonical variables are not justified by telling stories about thought experiments with the mythical Heisenberg microscope, but by rigorous deduction from the relevant axioms and definitions — a job that takes only two theorems, one definition, and a lemma borrowed from mathematics (Bunge 1967a: 252–256).

In conclusion, if we wish to avoid apriorism, anthropocentrism, and dogmatism, we must adopt scientific realism and reason from principles, not opportunistically. In turn, principled reasoning in factual science requires replacing the obiter dicta of famous people with axiomatized or at least axiomatizable theories that take care of content as well as of form. And sketching the content of a theory starts by indicating its referent(s). Indeed, the least we are expected to know is what we are talking about: whether it is a thing out there, a mental process, a social fact, or a fictive object.

5.7 The Mental: Brain Process, Information Transfer, or Illusion?

At present, the two most popular opinions on the nature of the mind among the philosophers of the mind are informationism and materialism. The former holds that the mind is a stuff-free *information-processing device* — more precisely, a computer that runs on programs. By contrast, materialism holds that everything mental is cerebral. Another difference between the two views is that, whereas informationism is a promissory note, materialism is behind cognitive neuroscience, the most productive branch of psychological research as well as the scientific basis of biological psychiatry (Bunge 1987; 2010). Let us see briefly how the axiomatic approach may help evaluate both schools.

Let us start with informationism, the only precise formulation of which is the thesis that minds, like computers, are basically Turing machines. Let us quickly review the main traits of any axiomatic theory of these artifacts. The basic or primitive concepts are M (the set of Turing machines), S (the set of possible states of an arbitrary member of M), E (the set of admissible inputs to any member of M), and $T: S \times E \rightarrow S$, the function that takes every <state s, stimulus e> pair into another state t of the same machine, that is, such that $T(s,e) = t$.

Note that S and E are sets, that is, *closed* collections of items given once and for all, hence quite different from the variable collections found in living beings. Second, a member of M will jump from *off* to *on* only if it receives a suitable stimulus: it lacks the self-starting or spontaneous ability of human brains. Third, the machine goes from one state to another only if it admits one of the stimuli in E: it does not recognize novel stimuli and it does not come up with new states — it is neither adaptive nor creative.

In particular, the machine neither develops nor belongs to an evolutionary lineage. Whatever novelty it encounters is a product of the engineer in charge of it. In my pig Latin, *Nihil est in machina quod prius non fuerit in machinator*. Fourth, the Turing machine is universal: it does not depend on the kind of stuff or material, whereas human minds exist only in highly developed brains, which are altered by the ingestion of chemical agents, from coffee to LSD to heroin.

By contrast to machines, human brains can start spontaneously, are inventive or creative, distinguish concepts from their symbols (unless they are nominalists), enjoy some freedom or self-programming, and can invent problems whose solutions are not necessary for survival, such as "Could our omnipotent God hoist Himself by pulling His shoelaces?," which kept many a Byzantine theologian in business.

In sum, Turing machines behave as prescribed by the behaviorist psychology that ruled in the American psychological community between 1920 and 1960 — like well-trained rats and preverbal babies. On top of ignoring everything that distinguishes us from rats, that psychology rejects all attempts to investigate the neural mechanisms that explain behavior and mind. This is why it cannot help design treatments of mental diseases more complex and common than phobias, such as addiction, depression, and schizophrenia.

Contrary to information-processing psychology, cognitive and affective neuroscience makes precise and consequently testable hypotheses, such as "The cerebral amygdala is an organ of emotion," "Moral evaluations and decisions are made by the fronto-parietal lobe," and "The hippocampus is the organ of spatial orientation" — John O'Keefe corroboration of which was rewarded with the Nobel Prize.

These and other precise hypotheses can be cobbled together into psychoneural theories, such as the author's quasiaxiomatic one (Bunge 1980). This theory consists of 27 postulates, 16 theorems and corollaries, and 44 definitions, all of which refer to neural systems and their specific functions, that is, the processes peculiar to them.

For example, Definition 7.9 (iv) in that book reads thus: An animal *a* is *creative* = "*a* invents a behavior type or a construct, or discovers an event, before any other member of its species." Then follows Postulate 7.5, inspired in Donald Hebb's 1949 seminal work: "Every creative act is the activity, or an effect of the activity, of a newly formed neural system."

5.8 Axiomatic Theory of Solidarity

Since Ibn Khaldûn's seminal work in the 14th century, it has been known that solidarity, rooted in shared interests and values, is a mechanism of survival and social cohesion of human groups, starting with the family, the gang, and the village. In particular, marginal people survive because they practice mutual help, as Larissa Adler-Lomnitz (1975) showed in her pioneering study of Mexican shantytowns.

It is not for nothing that solidarity is a member of the most famous political slogan in history: *Liberté, égalité, fraternité*. However, there are few scientific studies of solidarity; worse, this same-level process is often confused with charity, which is top-down. In the following we present a mathematical model of solidarity, just to illustrate the dual axiomatics format. However, we start by presenting it heuristically.

We shall say that two individuals or social groups are mutually *solidary* if and only if they share some of their material resources, that is, if each of them hands over to the other a part of his/her concrete goods or bears some of his/her burdens.

One way of formalizing the solidarity concept is to assume that the rate of change dR_1/dt of the resources of unit g_1 is proportional to the sum of its own resources R_1 plus the excess of g_2's resources over g_1's. In obvious symbols,

$$dR_1/dt = k[R_1 + (R_2 - R_1)] = kR_2 \qquad (1a)$$

$$dR_2/dt = k[R_2 + (R_1 - R_2)] = kR_1, \qquad (1b)$$

where k is a constant with dimension T^{-1}. Dividing the first equation by the second and integrating, we get

$$R_1^2 - R_2^2 = c, \qquad (2)$$

where c is another dimensional constant. The graph of the preceding equation is an hyperbola on the $<R_1, R_2>$ plane: see Figure 5.1. Since resources are positive quantities, only the first quadrant is to be kept.

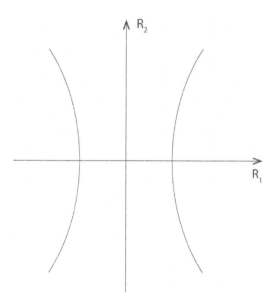

Fig. 5.1. Sharing resources: Graph of Eq. (2).

What follows is one of the possible dual axiomatizations of the preceding model of solidarity.

Presuppositions: Classical logic and elementary infinitesimal calculus.

Primitives: G, R_i (where $i = 1,2$), T, V.

Definition 1. The rate of change of resource R_i is dR_i/dt.

Axiom 1m G is a countable set.

Axiom 1s Each element g_i of G represents a social group.

Axiom 2m Every $R_{i,}$ with $i = 1,2$), is a function from G to the set V of real positive numbers, and differentiable with respect to t, where t is an arbitrary element of T.

Axiom 3s. R_i denotes the material resources of group g_i.

Axiom 4m. k and c are positive real numbers, the dimension of k is T^{-1}, and that of c is the same as that of V^2.

Definition 2. The social groups g_1 and g_2 are *solidary* to one another if and only if they satisfy conditions (1a) and (1b).

Corollary 1. Only the first of the quadrants in the $<R_1, R_2>$ plane has a sociological sense.

This is an immediate consequence of Axiom 2s since, by definition, all resources are positive. Physical parallels: the advanced waves in classsical electrodynamics and the negative probabilities in a now forgotten theory of Dirac's.

Theorem 1. The system of differential equations (1a) and (1b) implies the algebraic equation

$$R_1^2 - R_2^2 = c, \qquad (2)$$

where c is another dimensional constant. The graph of this equation is a vertical hyperbola on the $<R_1, R_2>$ plane.

Proof. Dividing (1a) by (1b) results in

$$dR_1/dR_2 = R_2/R_1$$

Integrating one obtains (2).

Theorem 2. Solidarity approaches equality as both the resources and solidarity increase.

Proof. The asymptote of the right branch of the hyperbola (2) is the straight line "$R_1 = R_2$", representing equal resources.

Remark. A political moral of Theorem 2 is that, since solidarity breeds equality regardless of the initial endowments, it is not necessary to impose it by force. However solidarity does not emerge by itself: it is but one of the sides of the triangle *Liberty, equality, solidarity*. Take note, Pol Pot's ghost.

5.9 Virtues of Dual Axiomatics

Axiomatization is required when the precision of some key concepts is questioned, or when the basic assumptions must be identified to ensure their covariance (or frame invariance). However, when either the denotation or the connotation of a theory is not clear, we need to state them explicitly, in which case we must engage in dual axiomatization. This boils down to enriching the formalism of a theory with a set of semantic assumptions, such as "General relativity is a theory of gravitation" — rather than, say, a theory of spacetime, or a generalization of special relativity.

The main virtues of dual axiomatics are the following:

a. *It preserves the form/content*, a priori/a posteriori, and rational/factual dualities;

b. *It unveils the tacit assumptions,* in particular the vague or false ones, which is where the dog was buried, as a German might put it;

c. *It reminds one from start to finish which are the referents* of the theory, which prevents philosophical and ideological contraband; for example, it shows that the presentations of the relativistic and quantum theories in terms of measurements are false, since the concepts of observer, apparatus and indicator do not occur in the principles;

d. *It helps disqualify extravagances* such as the many-worlds and branching-universes fantasies, as well as John A. Wheeler's thesis that *its* (material things) are nothing but bundles of *bits* (information units), by showing that they violate standard conservation principles;

e. *It exhibits the legitimate components of the theory as well as their deductive organization, and with it the logical status of each constituent* (universal/particular, primitive/defined);

f. *It facilitates the empirical test of theories,* in showing the absence of indicators or markers in it, hence the need to add at least one indicator for every type of measurement; and

g. *It eases the understanding and memorization of theories,* in highliting the most important constructs and thus reducing the number of formulas required to reconstruct theorems.

Coda

We conclude that dual axiomatics is anything but a dispensable luxury, for it helps to detect false presuppositions, gaps, weak links, and pseudotheorems; it is indispensable for doing serious, deep and useful philosophy of advanced science; and it suggests that deep science presupposes a pro-science philosophy.

CHAPTER 6

EXISTENCES[1]

The problem whether something is real or imaginary may be existential or academic. Suffice it to recall that the so-called sacred scriptures of the monotheistic religions mandate killing anyone denying or even doubting the existence of God; that claims to the existence of atoms, the aether and phlogiston, action at a distance and spontaneous generation, evolution, and genes, have split scientific communities for centuries; that the myth of the existence of the vaginal orgasm was exploded as recently as 1966; and that the possibility of free will is contested as vehemently as when Augustine asserted it 15 centuries ago.

Ordinary folks in good mental health do not doubt the real existence of their surroundings. Only philosophers can afford to deny it while having no doubts about their own existence — and importance. Common folks used to be locked up in nuthouses if they expressed such ontological nihilism.

But, of course, it is not the same to affirm the existence of something as to claim that it has such and such properties. For example, as a distinguished cosmologist wrote recently (Cirelli 2015), "Dark Matter exists, and discovering what it is made of is certainly one of the major open problems in particle physics and cosmology nowadays."

6.1 Introduction: It's Not ∃

A naïve reader is likely to think that existentialists will elucidate the word *existence*, and will consult Heidegger's main book, *Sein und Zeit* (1993 [1926]: 440). There he will be revealed that existence "is related to the truth of being itself." But this sentence won't help anyone curious of

---

[1]Revised version of the paper with the same title published in *Review of Metaphysics* Vol. 70, No. 2, Dec 2016.

whether quarks or ovnis exist, if only because the expression 'truth of being' makes as much sense as 'being of truth.'

Given the persistent confusions about existence in the literature, from Parmenides to Heidegger to dark matter hunters, it won't harm to repeat Hamlet's most famous saying. The lowest of insects admits tacitly that Hamlet's dilemma summarizes the struggle for life, though it might rephrased as "To eat or to be eaten."

One of the most prestigious of the confusions about existence is the belief that the so-called existential quantifier $\exists$ exactifies the notion of existence in all fields. That this is a plain mistake, is realized upon recalling the way $\exists$ is defined, namely as *not-all-not*, or $\exists xPx = \neg\forall x\neg Px$. Indeed, this formula should be read "Some individuals lack the property P," as in "Not everyone is a non-smoker." In sum, $\exists = some$, not *there is* (Bunge 2012).

In other words, with all due reverence for Charles Sanders Peirce, Bertrand Russell, Van Quine, and Alfred Tarski, $\exists$ should be rechristened *the someness quantifier*. The immediate reward for admitting this correction is that it eliminates Quine's question of the "ontological commitment" of logic. Logic has no such commitment, for it is *de dicto*, not *de re*. As Leibniz wrote, its truths (the tautologies) are *vérités de raison*, not *vérités de fait*.

Such topic indifference is why logic can be used everywhere. That is also why Hegel's notion of a dialectical logic is nonsensical. Conflict or competition is as ubiquitous as cooperation, but contradiction proper is about propositions, not existents.

6.2 Real Existence: Concept and Criterion

Let us now tackle real existence, the concept occurring, for instance, in the recent doubts about the claim that the Higgs boson was discovered at CERN in 2012. We shall distinguish the *definitions* of this concept from the *criteria* for finding out whether or not something exists really, or is in the world. Whereas a definition of a concept answers the *What is it?* question, a real existence criterion answers the *How do we know?* question. Whereas the latter is epistemological, the former is ontological. Gold is the element with atomic number 79, and it may easily be recognized by dropping a bit of *aqua regia* on it, without resorting to either authority or argument.

In line with my materialist or realist ontology (Bunge 1977; 1979b), I propose the following.

Definition 1: **Real existence = materiality = mutability**

To put it formally, for all x: x *exists really* = x is mutable. If preferred, $\forall x$ (x is an existent = x is capable of changing).

Note that, following Alessandro Padoa's advice, to define we use identity (=), not the much weaker equivalence relation (iff). Thus, for all x, x is alive iff x metabolizes, but there is much more to life than metabolism (i.e., life ≠ metabolism). Note also that this type of existence is absolute or context-free. In particular, it does not depend on human experience.

Since in principle every existent x can be ascribed at least one state space $S_r(x)$, or set of all possible states of existent x relative to a reference frame r, the above definition may be replaced with:

Definition 2: **An object x exists really = Every state space $S_r(x)$ for x has at least two elements**

For example, if a and b name two different possible states of x, such as

$a = x$ is at place p relative to frame r at some time t_1, and
$b = x$ is at place q at time t_2, where $p \neq q$,

then x may be involved in two different events during $[t_1, t_2]$:

$<p, r, t_1> \rightarrow <q, r, t_2>$, and $<q, r, t_1> \rightarrow <p, r, t_2>$.

Consequently, x exists really during the time interval $[t_1, t_2]$ = x is an existent over $[t_1, t_2]$.

Finally, we stipulate the following real existence criterion or indicator:

Criterion 1. Individual x *exists really* if and only if x makes a difference to at least one other existent.

More precisely, for all x: x exists really relative to frame r and at time t if and only if $\exists y\{(y \neq x)\ \&\ [S_r(y) \neq S_r(x)]\}$, where $S_r(x)$, $S_r(y) \neq \varnothing$.

Equivalently, x exists really relative to frame r and at time t if and only if x *acts* upon y or conversely. In symbols, $A_{r,t}(x,y) = [S_r(y)\ \Delta\ S_r(x)]$, where Δ stands for the difference between two sets. That is, $A\Delta B = (A–B) \cup (B–A)$ = everything in A but not in B plus everything in B but not in A.

Finally, note that *real existence is absolute*. In particular, it does not depend on human experience: the above definitions and criteria are not egocentric. By contrast, subject-dependent existence can be characterized by

***Definition 3:* An object x *exists phenomenally* = x occurs in someone's sensory experience**

More precisely: for all *x*: *x exists phenomenally* if there is at least one sentient being that feels *x*.

Note, firstly, that, unlike real existence, phenomenal existence is relative to some sentient subject — whence it may also be called *subjective*. Secondly, the subject in question is any organism capable of sensing external stimuli. Thus even the lovely *Mimosa pudica* weed, whose leaves fold when touched, can be said to detect phenomenal existents. This well-known fact raises the question whether phenomenalist philosophers like Hume, Kant, Mach, Carnap, and Goodman, should be lumped together with sensitive plants.

6.3 Conceptual Existence

Conceptual existence is occurrence in a conceptual system, that is, a collection of constructs held together by a binding relation such as concatenation, implication, addition, function, or morphism.

In short, we propose

***Definition 4:* S = <C, •> is a conceptual system = C designates a set of constructs, and • a binary relation in C**

Obvious examples of conceptual systems are propositions, graphs, groups, categories, classifications, and theories (= hypothetic-deductive systems). By contrast, sentences are not systems unless their key terms are interpreted, or assigned meanings, and thus converted into the linguistic counterparts of propositions.

We are now ready for

***Definition 5:* For all x, x *exists conceptually* = x is a constituent of a conceptual system**

For example, p∨¬p exists in the system <L, ∨,∧,¬> of classical tautologies, but not in that of intuitionist logical truths. And the number √2 exists in the system <ℝ, +, · , ⁻¹, < > of real numbers, but not in the algebra of classes or in Peano's system for natural numbers.

The mathematical existence (and non-existence) theorems constitute the purest specimens of conceptual existence. Let us briefly recall two of them: the irrationality of $\sqrt{2}$ and Fermat's last theorem. The earliest existence (or rather non-existence) theorem was perhaps the statement that *there are no* two positive integers *m, n* such that their ratio *m/n* equals $\sqrt{2}$. An equivalent statement is that no positive integers *m, n satisfy* the equation "$\sqrt{2}$ = m/n." Shorter: "$\sqrt{2}$ is an irrational number." Likewise, Fermat's last theorem states that no three positive integers *a*, *b*, and *c* satisfy the equation $a^n + b^n = c^n$ for any integer value of *n* greater than 2.

In both cases, the existence of an abstract object has been replaced with its *satisfaction* of a formula in some model or interpreted theory. This kind of existence is thus *relative*, by contrast with the existence of, say, the sun, which is *absolute* in that it does not depend on anything else. (Note that in a logically consistent monotheistic theology only God exists absolutely.)

No such substitution is possible in the factual sciences and technologies, where (real) existence (or non-existence) is absolute. For example, asserting that perpetual motion machines are impossible is not quite the same as saying that such an engine machine *would* violate (or fail to satisfy) the first law of thermodynamics. Indeed, whereas the first statement has only one referent, the second has two and, moreover, it is a counterfactual. And counterfactuals, the darlings of possible-worlds fantasists, are not admitted in scientific or technological discourses except as heuristic devices.

Besides, a radical skeptic might argue that the first law is just a hypothesis, so we should not disqualify a priori any research on perpetual motion devices. Fortunately, since the mid-19th century neither physicists nor engineers have wasted their time attempting to refute the said law.

Neither the conceptual existence concept nor its dual is replaceable in the majority of mathematical existence theorems. Think, e.g., of the intermediate value theorem, which asserts the existence of a point ξ, in the [a,b] interval of the horizontal axis, where a continuous function *f*, such that $f(a) > 0$, and $f(b) < 0$, vanishes, i.e., $f(\xi) = 0$.

But for this theorem, a material point could not move smoothly from the first quadrant to the fourth. The radical constructivists (or intuitionists) refuse to accept this theorem because it does not tell us how to construct the functions that satisfy it. Let them pay for the loss of that wonderful theorem.

However, the most hotly contested existence statement in the history of mathematics is the *axiom of choice*, usually attributed to Ernst Zermelo but actually anticipated by Giuseppe Peano and Beppo Levi. Roughly, this axiom states that, given a possibly infinite family of non-empty disjoint sets, there is a function, called the *choice function*, that picks one element of each set. The domain of this function may be pictured as the collection of electoral districts of a country, and its codomain as the parliament of their representatives.

Constructivists object that this axiom does not specify how to construct the choice set. Everyone else accepts the axiom. The Platonists because it has been proved that set theory is consistent with or without that axiom. And the rest accept the axiom because it "works," in that it is used to prove theorems in several branches of mathematics.

The axiom of choice is firmly entrenched in the body of mathematics. Indeed, it is equivalent to several other key mathematical statements that at first sight are alien to it. One of them is Zorn's lemma, which reads thus (Halmos 1960: 62): "if X is a non-empty partially ordered set such that every chain in X has an upper bound, then X contains a maximal element." For example, if $A = \{a,b,c\} \subseteq X$, and $a < b < c$, then there is a u in X such that, for every x in X, if $u \le x$, then $u = x$.

From the fictionist viewpoint (Bunge 1985a; 1994; 2012a), the debate over constructivity is a storm in a teapot. Indeed, whether or not there is a constructive proof of a given mathematical object, this is just as fictitious as Zeus or as a talking dog. Unlike abstraction, fictiveness does not come in degrees any more than real existence does. Only those who, like the nominalists, fail to distinguish conceptual from material existence, can get excited over the debate in question.

6.4 Semiotic Existence

Driving down a road I see a stop sign, and I immediately press the brake pedal. Should the stop sign be attributed existence? Undoubtedly, since I reacted to my perception of it. The sign in question has what may be called *semiotic existence*, or *existence by proxy*.

Of course, the road sign does nothing by itself, but my reading and understanding it has a causal power, hence it must be attributed real existence, which it lacks to someone who has no inkling of the language it is written it. The causal chain is: light beam reflected by the roadsign → my cognitive system → my voluntary action system in my prefrontal cortex → my right leg-and-foot system → brake pedal → my car's braking system → my car's slowing down.

The preceding suggests the following

Definition 6: **The object** *x exists semiotically* **= some animal ψ is capable of producing reaction** *z* **upon perceiving and evaluating** *x*

A second type of semiotic existence is what may be called *denotational reality*, as in "Contrary to conventionalism, the field equations are not just computational tools but represent physical entities." This suggests

Definition 7: **The symbol** *S* **is** *realistic* **(or** *exists semiotically***) = there is a real existent denoted by** *S*

This concept occurs implicitly in the discussions, still going on, about three important physical symbols: the electrodynamic potentials $A_{\mu\nu}$, the metric tensor $g_{\mu\nu}$ in the theory of gravitation, and the state function y in quantum mechanics. It may be argued (e.g., Bunge 1987; 2015) that all three symbols are endowed with physical meanings: the first two denote fields (the electromagnetic and gravitational ones, respectively), and the third denotes quantum-theoretical entities.

6.5 Fantastic Existence

Fantasies can be said to exist in their own contexts. More precisely, we propose

Definition 8: **For all** *x*: *x exists fantastically* **= there is a work of fiction that contains or suggests** *x*

For example, Shakespeare's Caliban exists, or "makes sense," in *The Tempest*, but nowhere else. The same holds for the myriads of Hindu divinites: their worshippers reify them by flinging lumps of butter at their images. The previous sentence evokes Jorge Luis Borges's assertion that theology is the most perfect specimen of fantastic literature. Actually all literature proper, unlike weather reporting and honest accounting, is fantastic to some extent, which is why we read it: not to learn something, but to be moved or uplifted, challenged or entertained.

The same holds for music, the plastic arts, and artistic cinema: all their specimens, even the films of the Italian realistic school, are sheer fantasies. And fantasy comes in degrees. Thus, Italo Calvino's nonexistent knight is even more fantastic than his cloven viscount; and abstract mathematics is further removed from reality than either geometry or number theory.

In both the cases of artistic experience and religious worship we let ourselves be overwhelmed by fiction and detachment from reality. Thus, immersion in either art or religion involves the involuntary denial of reality — the seal and test of voluntary temporary insanity.

We fantasize some of the time in all walks of life, sometimes to escape from reality, and others to cope with it. In the famous Italian film "Pane, amore e fantasia" (1953), a ragged man lunches on a loaf of bread seasoned only with love and fantasy. By contrast, Gina Lollobrigida, whom that film gave instant celebrity, was abundantly real.

Mathematicians and theoretical physicists are professional fantasizers. But their fantasies, unlike those of Maurits Escher, are bound by reason. In fact, mathematical activity consists most of the time in proving theorems — that is, in forcing certain items into pre-existing conceptual systems. And, as David Hilbert remarked a century ago, theoretical physicists have a harder time than pure mathematicians, for they are expected to justify their inventions in terms of empirical findings. Indeed, when their fantasies turn out to be wild, like those of string theorists and many-worlds fans, they are rightly accused of perpetrating pseudoscience (see, e.g., Smolin 2006).

According to Plato's Socrates, the unexamined life is not worth living. (Kurt Vonnegut commented: "But what if the examined life turns out to be a clunker as well?") Much the same may be said about life without fantasy, since it takes a lot of fantasy to conceive of new theories and new artifacts, as well as to design new feasible courses of action, and even to estimate their possible moral values.

6.6 Surrealism

The quantum theory was initially so counter-intuitive, that it was interpreted in many different ways. Even today, after nearly one century, many interpretations of it coexist, even though most textbooks adopt the so-called Copenhagen interpretation proposed by Bohr and Heisenberg.

This interpretation is operationist, since it revolves around the concept of an observer even in the case of an atom situated in the center of a star. So, it is irrealist, but at least it is not insane. By contrast, other interpretations of the same theory are utterly absurd. One of them is the so-called many-worlds interpretation proposed by Hugh Everett (1957) at the suggestion of his thesis advisor J.A. Wheeler, the author of a number of additional weird ideas, such as the "its-from-bits" fantasy.

The kernel of Everett's interpretation is that every time we perform a measurement we choose one possibility without thereby killing the remaining possibilities, which are realized in alternative worlds. For example, Schrödinger's vampire cat, which was half-alive and half-dead while locked up in its cage, may jump out of it when the observer opens it; but the cat's corpse, far from vanishing, takes up residence in an alternate world. How do we know that this is indeed the case? Simple: Everett *dixit*. What becomes of the actual/potential split, and where does the extra energy for the multiplication of worlds come from? Ah, only one story at a time, please!

Physical surrealism is not confined to quantics: it is equally fashionable in cosmology, where it is usually called the *multiverse* hypothesis. The main objection to all the theories that postulate the actual existence of a multitude of parallel universes is that these are assumed to be inaccessible from ours, so that the said theories are in principle untestable, hence unscientific. Another objection is suggested by Ockham's advice not to multiply unnecessaily the number of entities.

Coda

We have argued that there are existences of five types, only one of which — real existence — is absolute, that is, context-independent, in particular subject-free. How similar are the various existences, and how are they related to one another? Let us see.

a. Real existence is absolute or unconditional, hence it must be either postulated or proved experimentally. Furthermore, real existence does not come in degrees: For all x, x either exists really or not. The concept of partial existence, about which Jacques Maritain wrote, is a theological fiction only necessary to make sense of the assertion that God is the *ens realissimus* — the uppermost link in the Great Chain of Being.

b. Phenomenal existence is relative, for it occurs only in sensory apparata, whether rudimentary like a worm's, or highly developed like ours. Let us not tell a schizophrenic that the monsters he claims to see or feel "are only in his mind," for he perceives them vividly and often painfully as well, as lurking out there. After all, the organ of phenomena, namely the nervous system, is objectively real. Thus, phenomena may be said to be only once-removed from objectively real processes.

c. Conceptual existence is relative to some conceptual system or other, within which it must be either assumed or proved exclusively with conceptual resources. For example, $\exists$ is either defined in terms of $\forall$ and $\neg$, or introduced via some postulates of the predicate calculus. And mathematics as a whole depends on the work of mathematicians, who are of course real entities. Hence, even the most abstract concepts presuppose the real existence of abstractors. In general, all concepts, even those that do not refer explicitly to facts, are *fact-coordinated*, as Rescher (1985) put it. In particular, the central concepts of philosophy ("mind," "truth," "good," etc.) "are importations from everyday life and from science" (Rescher op. cit.: 45).

d. Semiotic existence is attributable only to signs. And these are perceptible objects like numerals, hence objectively real as well — though only an educated perceiver can endow them with signification. Hence after a nuclear holocaust the ashes of books would be just physical things rather than semiotic ones.

e. Fantastic existence occurs only in works of fiction while being read by people capable of fantasizing. Hence fantastic existence escapes literal minds.

f. The general concept of contextual existence can be introduced by the following convention, suggested in an earlier publication (Bunge 1977):

Definition 9: **Let** U **designate a well-defined universe of discourse or collection of objects, and call** χ_U **the characteristic function of** U, **defined by this pair of value assignments:** $\chi_U(x) = 1$ **iff** x **is in** U, **and** $\chi_U(x) = 0$ **iff otherwise.** The *existence predicate* is the function E_U from U to the set of existential propositions, such that $E_U(x) = [\chi_U(x) = 1]$. If U is a collection of real items, then the existence in question is real, it is semiotic if U is an assemblage of signs, and so on.

The above definition suggests the invalidity of the once-famous assertion that "existence is not a predicate." Claims to existence and its dual are too important to be admitted or rejected without evidence, and existence of any kind is too serious to be left to existentialists, the arch-enemies of everything scientific.

Chapter 7

REALITY CHECKS

Usually we solve daily life problems by resorting to either custom or authority: the subordinate consults his supervisor, the sick person her physician, the teacher her textbooks, the believer her confessor, and so on. Only research scientists, innovating technologists and the like check their principles and rules before applying them.

In particular, scientists check their guesses, or at least they hope that someone else in the scientific community will eventually subject them to rigorous tests. True, they too use the argument from authority every time they rely on a scientific table or paper. But such texts are expected to be trustworthy and replicable rather than holy words spoken from a hill.

Moreover, the users of such sources will correct them whenever an error in them is discovered. And the lazy thesis supervisor will suggest replicating a celebrated experiment hoping to catch a significant error hiding in it. In short, the authority in question is regarded as transient and disputable.

Incidentally, replication is the rage in the so-called 'soft' sciences at the time of this writing, after an unusually large number of papers had to be retracted for failing the replication test. Some people believe that those disciplines are going through a replicability crisis.

In sum, scientists and innovative technologists are expected to check their assumptions and conclusions. Let us peek at some of the philosophical issues arising from such checks, starting with the idea of evidence, which is anything but self-evident.

7.1 Facts, Data, and Peta

In this section we shall discuss the way our ideas about facts are checked for truth. Let us start by agreeing on the meanings of the polysemic term

91

'fact.' We shall mean by 'fact' either the *state* of a thing, an *event* involving it, or a *process* it is undergoing. For instance, that this glass contains an ice cube is is a fact; that the ice has started to melt is a fact of the event kind; and that it underwent a transition from the solid to the liquid states is a fact of the process kind. Note that in all three cases a material thing was involved: all states are states of some concrete thing, and the same holds for events and processes. There are no states, events or processes in themselves: every fact involves things.

The above ideas can be formalized by using the concept of the *state space* $S(\theta, f)$ for a thing θ relative to a reference frame f at time t:

Thing θ is in state s relative to frame $f = s \in S(\theta, f, t)$

Event $e = <i, f> \in S(\theta, f, t) \times S(\theta, f, t)$ happens to θ relative to f at t.

Thing θ undergoes process $\pi = < s \in S(\theta, f, t) \ t \in \Delta t >$ relative to f over time interval Δt.

Now, hypotheses are checked for (factual) truth, and rules for efficiency. Both are said to be subject to *reality checks*, or to be *contrasted with reality* — a process that, of course, presupposes the realistic postulate.

However, such contrast cannot be literal, since only objects of the same kind can be compared with one another. We can compare two facts or two statements in some respects, but it is not clear how to compare a fact, such as a given rainfall, with a statement such as "It is raining." And saying that the former is a "truth maker" for the latter is just verbal juggling. Indeed, a fact can alter or generate other facts, but it cannot assign truth-values: only people can perform such evaluations.

And the latter bear on propositions or their linguistic wrappings, namely sentences in some language.

We may also say that there can be *evidence* for or against the statement that it is raining here and now: we see rain falling, wetting things, and gathering in puddles on the ground, we can touch it when going outdoors, etc. So, *we confront the statement "It rains" with the relevant data*, and conclude either that the statement in question has been verified, confirmed or corroborated, or that it has been falsified, infirmed or refuted.

Equivalently, we conclude that the data in hand constitute *evidence* for the statement about rain. In other words, we have gone through this process: *Hypothesis statement — Data collection — Confrontation of hypothesis with data — Hypothesis evaluation.*

So much for *data* or the given, that is, perceptible facts — the ones that empiricists care for. Such data are indispensable for most ordinary life pursuits, but insufficient in science, which deals mostly with occult factual items, such as electric fields, chemical reactions, mental processes, and the past. In doing scientific research we must *seek out* facts instead of waiting to get them for free. In short, in science we deal more with *peta*, or sought facts, than with data or givens.

How are peta produced? The modus operandi depends on both the facts of interest and our hunting gear. For example, to find out whether or not an electric cable is "live," we place a magnetic needle near it and remind ourselves of the theory about the ponderomotive forces exerted by the invisible magnetic field that accompanies an electric current; to reconstruct our remote hominim ancestors we dig for fossils and try to imagine how they lived; and to understand why we must pay taxes we remind ourselves that they were instituted to pay for public services.

Each of these *peta* was obtained on the strength of some hypothesis or theory, and it is evidence for the actual occurrence of a fact that, though imperceptible, is or used to be just as real as the fact(s) referred to by a datum or given. For example, since about 1850 several expeditions have been mounted to look for remains of our earliest ancestors. Only evolutionary biology justified such travails and expenses. The recent Chinese findings have suggested a new hypothesis about the most likely dispersal routes: see Figure 7.1.

7.2 Indicators

To get to know something about conjectured imperceptible facts we must connect them with perceptible facts, such as the height of a mercury column, the position of a pointer in an ammeter, the frequency of clicks of a Geiger counter, or the density of grains on a photographic emulsion exposed to cosmic radiation.

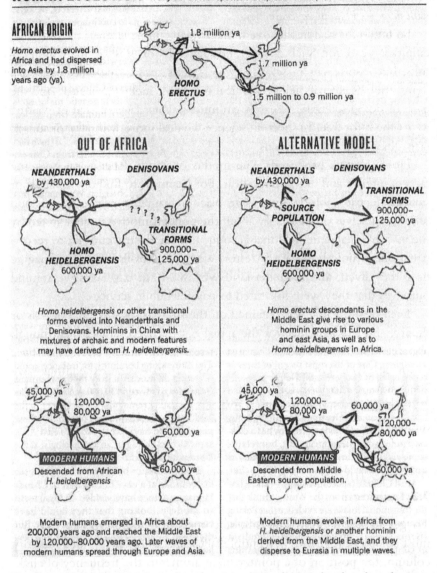

Fig. 7.1. Alternative hypotheses about hominim dispersal. Note that they share the guess that our species emerged in East Africa some 600,000 years ago. Reproduced from Qiu, Jane. 2016. "The forgotten continent," *Nature* 535: 218–220.

Furthermore, ideally such a connection should be lawful rather than arbitrary. That is, there ought to be a well-corroborated functional relation $p = f(i)$ between an imperceptible variable i, such as wind speed, and a perceptible marker or indicator p, such as the numeral representing the number of turns of the cups of an anemometer over a time interval. This number is lawfully related to the wind speed, a relation can be checked in a wind tunnel. An even simpler example is the stretching of the spring of a spring scale, which, to a first approximation, is proportional to the load (Hooke's law).

While this particular indicator is intuitive, most other scientific indicators are not. Think, for instance, of hyperglycemia as an indicator of pancreas malfunction, such as diabetes mellitus, via the causal chain: Pancreas malfunction → Insulin deficiency → Sugar excess → Sweet urine → Agglomeration of flies around a urine pool. Oscar Minkowski and Joseph von Mering established that chain in 1889, by studying the physiology of a dog surgically deprived of its pancreas. Incidentally, they solved the inverse problem Symptom → Disease, by transforming it into the direct problem of experimentally causing the disease by pancreas ablation, and observing the consequences. Why did Minkowski suspect the pancreas instead of another organ, is another story.

In short, to test a hypothesis about unseen facts of some kind we must craft a second hypothesis helping us device a perceptible indicator or marker of the imperceptible facts in question. Every indicator is specific, that is, it depends critically on the kinds of stuff both indicator and indicated are made. Thus, scales cannot be made with putty, nor can a Geiger counter or a pH meter be built wholly in wood. Hence the very idea that there are universal or stuff-free indicators and instruments is false. And yet this wrong assumption is common to all the so-called general measurement theories, such as von Neumann's, as well as to the empiricist texts on empirical operations, all of which assume that all measurements are direct. Moreover, the very notion of an indicator fails to occur in the vast majority of texts on the philosophy of science.

Worse, there is a whole school in social studies whose only postulate is that such studies *must* dispense with the scientific method because they consist in "interpreting" facts — and of course interpretations, unlike scientific hypotheses, are subjective and therefore incomparable with one another.

7.3 Theoretical Models

No general theory, such as quantum mechanics, the theory of evolution, or rational choice economics, can directly account for particular facts without further ado. Every application of a general theory G to facts of a particular kind requires enriching G with a set P of particular data or peta about the things to be accounted for, such as their number, density, electrical conductivity, occupation, or income bracket.

For example, there is a theory of planetary motions, and a very different one of bicycle riding; we should use one theory of each of the hundred plus chemical elements to account for their different spectra; and, pace received economic wisdom (or folly), we should not expect a single economic theory to fit both the American and the Honduran economies.

However, most theoretical models in social science are built from scratch, without leaning on general theories. Gerd Buchdahl called them *free*; I prefer to call them *ad hoc*. Braithwaite called *teoritas* (Spanish for little theories) both free and bound theoretical models, for their reference classes are narrow. In any event, they should not be confused with the models studied in mathematical model theory, since these are examples of abstract theories. For example, an arbitrary set together with an associative operation constitues a semigroup; a model of a semigroup is the set of integers together with +. The structuralists confuse the theoretical models in factual science with the models in model theory.

Not only stuff matters in science: structure and environment too matter. For example, two chemical compounds may be composed of the same atoms but, if their structures are different, their global properties too will differ: they are isomers of one another. And the molecular differences between isomers are bound to have counterparts at the macrolevel. Just think of the enormous biological differences that result from different orders of the nucleotides A, C, G, and T. No wonder that biological markers, such as pulse, are so different from molecular markers, such as acidity or chirality.

Most philosophers have ignored or minimized the role of theory in the design of measurement instruments and experiments, as well as in the evaluation of scientific theories: this is why empiricism has utterly failed to account for scientific experience. Let us briefly evoke two examples: the genesis of special relativity and of X-ray crystallography.

Einstein's special relativity (*SR*) was born from a comparison between two theories about things of very different kinds, namely classical mechanics, which is about bodies, and classical electrodynamics, which concerns electromagnetic fields and their sources. When Einstein started thinking about their relations, he knew that his seniors Henri Poincaré and H.A. Lorentz had noted that, whereas Newton's laws of motion of bodies were invariant under the frame transformations in the Galileo group, Maxwell's did not change under transformations belonging to a more comprehensive group, namely Lorentz's.

Einstein's original contributions were to postulate that (a) this duality had to be eliminated, (b) such elimination required choosing one of the theories as being more powerful (truer and deeper) than the other, (c) preferring electrodynamics over mechanics, and (d) reconstructing mechanics so that its laws of motion were invariant with respect to the Lorentz transformation.

The outcome altered mechanics in a manner that shocked all those who believed that rational mechanics was just an application of mathematics, hence immune to experiment; it also shocked the empiricists, for Einstein's formulas did not emerge from any measurements — in particular, they were not condensates of the Michelson and Morley exquisitely precise measurements. (Ironically, neither of these men accepted Einstein's feat.)

Contrary to most expectations, Einstein's formulas for distance "contraction", time "dilation," and the increase of mass with velocity were experimentally confirmed, even by physicists who had tried hard to refute them. The philosophical moral, that sometimes theory could better old and seemingly self-evident empirical results, left empiricists and neo-Kantians speechless.

Even after one century, the scientific and philosophical literatures about *SR* are filled with serious errors, notably the confusions of relativity with subjectivity, and of invariance with objectivity; the belief that invariance principles are natural laws (rather than metalaws, or laws about laws); the belief that "$E = mc^2$" is universal, while in fact it does not hold for massless entities such as photons; and Paul Feyerabend's contention that the relativistic mass concept is "incommensurable" (incomparable) with the classical one (see Bunge 1967a, 1974a).

Background   Hypothesis
                              > Testable hyp. → Test → Raw data → Analysis → Conclusion
Worldview   > Indicator

Fig. 7.2.   Sketch of the present view of a scientific research project.

In short, modern science does not fit in with the standard conception of the scientific method, according to which it boils down to the sequence

$$Observation \rightarrow Hypothesis \rightarrow Test \rightarrow Data \rightarrow Analysis \rightarrow Conclusion$$

I submit that actual scientific research projects match the sequence sketched in Figure 7.2, showing that ideas precede observations.

7.4. Philosophy in the Lab: From Empiricism to Realism

Empiricism is still the prevailing philosophy of knowledge in the scientific community. From antiquity to the present, empiricism has been fashioned around such ostensible traits as shape, size, weight, color, and texture. It should have been obvious that this primitive methodology could not posssibly account for the scientific novelties of modern times, such as theory-guided experiments and the idealizations inherent in scientific models, such as the balls-and-rods models of molecules, and the neglect of the "imperfections" of the things modelled, such as local deformations and inhomogeneities.

To account for both the complexities of the real world and the simplifications inherent in all models we invent to account for real things, we must craft a theory of knowledge far more sophisticated than empiricism, whether classical or logical. This new epistemology is *scientific* (as opposed to naïve) *realism* (see Mahner 2001). Let us see how it works in the case of X-ray crystallography, the experimental method employed by Francis Crick, James Watson, Linus Pauling, and Rosalind Franklin in starting molecular biology.

If a beam of X-rays is aimed at a crystal, and the impacts of the waves diffracted by it on a photographic plate are recorded, one obtains either a set of concentric rings, or a set of parallel bands, neither of which resembles the arrangement of atoms or molecules in the crystal.

This problem is best tackled by transforming the given inverse problem, of guessing the said invisible arrangement from the visible diffraction pattern, into a bunch of direct problems, as discussed elsewhere (Bunge 2006). In fact, the X-ray crystallographer tries out the various possible crystal structures, and calculates the corresponding diffraction patterns produced by irradiating the conceptual crystal with an imaginary beam of X-rays. That is, s/he performs a Fourier analysis of each of the crystal types, compares the mathematical patterns with the real one, and chooses the one that best matches the real one.

This roundabout method is a sort of highly sophisticated form of trial and error. The said mathematical analysis had been invented a century earlier by Joseph Fourier, and was put to work by William Henry Bragg and his son William Lawrence, both of whom earned Nobel Prizes for their feat — an unexpected practical application of a beautiful piece of pure mathematics.

The experimental setup is roughly as follows. An X-ray beam strikes a crystal of unknown composition and structure, and the emergent rays strike a photographic plate. Finally, the visible concentric rings or parallel bands are compared with the hypothetical crystals, to see which of these theoretical models best fits the resulting picture. (See Figure 7.3.)

7.5 Induction, Deduction, or Abduction?

If someone wonders whether the inferential process sketched above is inductive or hypothetic-deductive, we answer that is neither. Indeed, the essential step in that inferential chain is the Braggs's method, which includes *inventing* a set of hypothetical crystal structures, *deducing* the corresponding hypothetical diffraction patterns, and *comparing* these with the real figure preserved on film (see Figure 7.3). Neither invention nor comparison is

LABORATORY *X-rays beam → Real crystal → Real diffraction figure*

↑↓ *Comparison & choice*

MATH. MODELS *Beam → Hypoth. crystals → Hypoth. diffraction figures.*

Fig. 7.3. X-ray crystallography.

either inductive or deductive. Incidentally, neither probability nor induction proper (jumping from particulars to a generality), the two poles of the logical empiricist (in particular Carnap's) account of science, occurs in the given inferential process.

As for Popper's falsifiability, we have already seen (Chapter 2) that it is not the seal of science. Admittedly, whether in mid-stream or in the final evaluation step, one always looks for possible counter-examples (exceptions). But there is no weeding without initial seeding, which is when invention must occur for a research process to be productive, original, and interesting.

Despite the increasing use of sophisticated techniques, from brain imaging to computer simulation to statistical analysis, only 39% of the prominent psychological papers analyzed by a group intent on enforcing replicability survived replication (Bohannon 2015, Open Science Collaboration 2015). Worse, about 75% of the papers on biomedical research have turned to be plain wrong (Ioannidis 2005).

Ioannidis (op. cit.) holds that some of the main sources of this problem are small sample size, failure to adhere to common design standards, "hotness" of the topic, and financial interests. To these we may add deference to authority and the pressure to publish, as well as the unconscious wish to confirm our prejudices, the jump from correlation to causation, and the special cognitive traps studied by Daniel Kahneman (2011) and his coworkers.

Whether the inclusion of courses on logic and philosophy of science in the science and medicine curricula would significantly improve the quality of scientific production, is debatable. Indeed, the coursework would presumably include reading Carnap's (1936) celebrated if absurd paper on testability and meaning, which confused methodology — the study of scientific tests — with meaning, a subject belonging to semantics. The course bibliography would also include one of Popper's failed attempts to elucidate the truth-value of a hypothesis in terms of its improbability, as though it made sense to gamble with truth. A dose of common sense might have avoided both mistakes and more. Medical treatments should be analyzed and subjected to reality checks before being prescribed.

7.6 Evidence-Based Philosophy?

Galileo and other makers of the Scientific Revolution had to waste some of their time battling the ruling philosophy, in particular the ossified Aristotelianism of the popular Cesare Cremonini, a colleague of Galileo's, who called him Simplicius in his famous dialogue on the two "systems of the world" (actually models of the solar system).

Most historians of philosophy have interpreted this episode as the decisive victory of empiricism over apriorism. They forgot that in Galileo's time empiricism, far from being original, was the establishment epistemology. Indeed, the central dogma of scholasticism was *Nihil est in intellectu quod prius non fuerit in sensu* ("Nothing is in the understanding that was not earlier in the senses").

The new science repeatedly violated this principle, for it embraced or introduced ideas, such as those of planetary orbit around the sun, inertia, the irrelevance of weight to the acceleration of free-falling bodies, and the primacy of primary over secondary qualities, that were inconsistent with observational data. In short, the Scientific Revolution did not endorse empiricism, but practiced a sort of synthesis of empiricism with rationalism, which Newtonianism and Darwinianism reinforced. Just think of the atomic doctrine, the circulation of the blood, and the theoretical discoveries of previouly unseen celestial objects, such as Neptune and black holes.

Everyone knows that the new science knocked down many a popular philosophical myth. What is still under debate is whether philosophy should continue to be an occasion for wild speculation, or should seek scientific evidence instead. This question rose when the invention of non-Euclidean geometries around 1800 confuted Kant's dogma that Euclidean geometry was an inborn doctrine, and moreover one that rendered experience possible.

Some neo-Kantians, particularly the learned Ernst Cassirer, attempted to patch up Kant's philosophy, so as to render it acceptable to scientists. But, like Gogol's much-darned greatcoat, Kantianism proved to be beyond repair. A similar fate awaited the logical positivists — in particular Rudolf Carnap, Philipp Frank, Hans Reichenbach, and Carl Hempel — who tried to revive positivism with a dose of modern logic. They too failed because they retained the phenomenalism of Kant and Mach. They thought that epistemology needs no ontological foundation, and should avoid all the

Big Questions, starting with those of the independent existence of the external world, the mind-body connection, and the search for social justice. In short, the neo-positivists were scientific philosophers only in name. But they were the only ones worth arguing with in their time, because they embraced logic and scientism.

In conclusion, evidence-based philosophy is still a tantalizing research project, to which this author has made some contributions (Bunge 1974–1989).

CHAPTER 8

REALISMS

Many of the most spirited and protracted controversies in modern intellectual history concern the reality of certain entities. Let this sample suffice: Do supernatural entities exist? Is there life after death? Do space and time exist out there and by themselves? Do atoms and the aether exist? Do chance and force fields exist? Are there Schrödinger cats? Are the strings and membranes of string theory for real? Are there centrifugal and actions at a distance forces? Are there alternative worlds? Did the Big Bang and abio genesis occur? Are biological species real? Are there goal-directed biological items? Are there biological laws? Are there talented and criminal genes? Do innate knowledge and telepathy occur? Do the self and free will exist? Do concepts exist, or only signs? Do the Oedipus complex and collective memory exist? Has socialism ever been practiced? Do rational egoism and collective rationality exist? Is anthropogenic climate change for real?

8.1 No Science Without Facts and Factual Truths

The vast majority of scientists take controversies about reality as parts of the ordinary process of knowledge acquisition, but they do not spend much time worrying about the concept of real or objective existence — or existence outside human minds. By contrast, scientists invest much ingenuity and time designing and operating instruments to find out whether something is real or imaginary. This is why they work in labs and observatories: to discover new things and events, or to check conjectures — in short, to evaluate reality claims. Eliminate the twin concepts of reality and factual truth, and you jump from science to science fiction or to fantastic art.

Nevertheless, some of the best scientists have made anti-realist declarations that are being repeated uncritically. For example, Bohr and Heisenberg have stated that physicists do not study nature but what we say about nature. But this is obviously false, and borrowed from some irrealist philosopher, not from their own scientific experience. For example, Bohr's atomic theory concerns atoms, not Bohr's statements about it; and atomic collisions occur out there, not in our linguistic apparatus. Likewise, Heisenberg's famous inequalities refer to "particles," not to epistemological or linguistic categories. Physical entities do not have grammatical properties, and statements cannot be accelerated by electric fields. Let us not be fooled by the philosophical remarks improvised by scientists who wish to show that they are up to date with academic philosophy.

In modern philosophy from Berkeley and Kant on, real things have been neglected, and so have the sophisticated procedures scientists have devised to check existential hypotheses — to the point that some of the most egregious nonsense has been dignified with the qualifier 'existential.' By contrast, many philosophers, even the self-styled positivists, regard the ontological notion of reality and the epistemological one of objectivity as problematic, to the point of often writing the words *fact* and *truth* between scare quotes. Let no one think that they are unsophisticated!

For example, Berkeley's sophistries against realism, and Kant's extravagant argument for the subjectivity of space and time, are still regarded as being more deserving of being taught than Eratosthenes's method to estimate the radius of Earth, Spallanzani's to find out how frogs reproduce, or the Whitehall studies, which proved that submission to undisputed authority sickens subordinates and shortens their lifespans.

In sum, whereas scientists have been studying reality, and technologists have altered it, the most prestigious philosophers since about 1700 have applied their considerable wits to doubting or even denying reality. The balance of this chapter is devoted to examining some of their arguments.

8.2 The Realist Thesis

Philosophical realism is the thesis that the world external to the subject exists independently of the latter. That is, the realists or objectivists, far from believing that they construct the universe as they perceive or think it, hold that it preexists them and that it now makes them, now undoes them. They add that the most we can achieve is to enrich the world with some artifacts, such as pens and schools, or impoverish it with others, such as guns and massacres. Both theses are shared by the realists of all hues, in particular the naïve ones, who believe that the world is like their neigborhood, and the scientific realists, who know that it requires a lot of research to discover some of what hides behind appearances, since these are subject-bound and hard to control.

Irrealism, by contrast, is the thesis that the universe depends on the subject who studies it. This egocentric opinion is shared by solipsists ("Only me"), constructivists ("Everything is a human construction"), phenomenalists ("Only appearances exist"), fans of the "participant universe" ("No world without subjects"), and believers that things are information bundles ("Its out of bits"). Egocentrism is also shared by those who repeat the vulgar opinion that "the world is the color of the crystal through which it is looked at," which up to a point holds for the social world since we make it, but is false of nature, which is colorless. For example, many poor citizens vote for plutocrats because they have fallen for right-wing demagoguery: they are led not by facts but what they make of facts, just as the Thomas "theorem" predicted.

Solipsism is the thesis that a tapeworm would hold if it could think: "I am alone and what surrounds me is my host. My motto is *Sum, ergo sunt.*" This thesis is so absurd, that no one in her right mind would seem to hold it. Yet, some scientific psychologists have writtten that "the brain is the organ that constructs the world", and even that "the world is an illusion caused by the brain." Let us hope that these weekend fantasists have got the suitable plans and materials, and that they will come up with better worlds than the imperfect one they share with the realists.

Of course these scientists cannot have meant what they wrote: they may have wished to say that human brains construct images or models of the world instead of photographing it. If so, why not say it clearly instead of imitating the constructivists who conflate cartographers with the demiurge, and their maps with the territories the latter intend to represent? However, the culprits may reply in their own defense that they have not found the words 'representation' and 'reference' in the most cited books on semantics, the theory of meaning and truth.

Let us next peek at the most widely diffused irrealist doctrines.

8.3 Phenomenalism and Phenomenology

The most popular version of irrealism is phenomenalism, the thesis that the constituents of the world are phenomena, that is, appearances to someone (Kant 1787: 724), or "the objects that may be known through experience" (Husserl 1913: 52). Yet both Kant and Husserl were so confused, that they denied having denying the autonomous existence of the universe.

Phenomenalism is a form of anthropocentrism, or at least zoocentrism, since without sentient beings there would be no appearances. To confute phenomenalism suffice it to recall that the beings capable of enjoying or enduring appearances emerged only a few billion years ago. In short, phenomena are the contemporaries of sentient cells. And it is likely that the animals that trusted appearances were short-lived and left few descendants. In short, evolution has favored realism.

The phenomenology of Edmund Husserl (1913), better called *egology*, retains Kant's thesis that the world is the totality of *Erlebnisse* or experiences. But unlike Kant, who professed to respect science although he did not understand it for his lack of mathematics, Husserl held that the external world is not worth exploring. In effect, following Augustine, Husserl enjoined us to examine ourselves, to become lost in thought, pretending that the world does not exist. This is why his philosophy is also called *egology*.

This operation of "putting the world in parenthesis" is called *épokhé*. This Greek word originally meant suspension of judgment, or retreat from commitment, as in "I do not claim to know whether napalm is good or bad for you." On various occasions Husserl himself (e.g., 1928: 31), stated categorically that, because of its inward-looking stance, phenomenology is "the utmost opposite of objective science." (See detailed criticisms in Bunge 1951.) Amen!

Egology is synonymous with egocentrism and escapism, and those who engage in *épokhé* are said to practice "phenomenological reductions." In everyday life they are said to be irresponsible, egoistic, undecided, or even foolish. For example, of someone who feigns not to have heard a legitimate request for help we may state, in a scholarly tone, that he is just practicing the phenomenological method. Or else, he may admit that his motto is that of the Italian blackshirts: "*Me ne frego,*" or, more delicately, "I could not care less."

8.4 Irrealism Is Recent and Inherent in Empiricism

It is often believed that realism is a marginal philosophy. Actually what has been marginal in Western philosophy over most of its history is irrealism. In effect, irrealism was unknown in antiquity and in the Middle Ages, on top of which its only contribution has been to divert philosophers from serious issues. This is why it is hard to find outside philosophy faculties, and why it has not participated in any of the philosophical controversies generated by modern science.

In fact, all the European philosophers before Berkeley (1710) were realists. They did not say it explicitly because there were no outstanding irrealists to argue with. The belief that the world is imaginary, and life is but a dream, is characteristic of a Buddhist school (Tola & Dragonetti 2008: 263) as well as of a few fiction writers, such as Calderón de la Barca, who sought to dazzle us with paradoxes instead of enlightening us with discoveries or inventions. Besides, the radical skeptics, like Sextus Empiricus and Francisco Sánches, have been and will always be exceptional, since nihilism hampers the struggle for life. Just think of those who denied the reality of the panthers that were chasing them.

Irrealism is an unexpected product of empiricism. I say 'unexpected' because the empiricist's intention is to stick to empirical data, understood as expressions of facts — whence the frequent confusion of facts with data. George Berkeley, perhaps the earliest consistent and radical empiricist, saw this clearly. This is why he had the nerve to claim that his stance, far from being paradoxical, was common-sensical, for it matches human experience, that is, our seeing, hearing, touching, tasting, and having propriocentive impressions. In other words, Berkeley's thesis was that there are only phenomena (*ontological phenomenalism*), so that only they can be known (*epistemological phenomenalism*). Lowly Worm might concur, but Lassie would bark.

8.5 Hermeneutics and Computationism

Hermeneutics, or linguistic idealism, is the doctrine according to which the world, or at least human society, is a language or "like" a language. In other words, hermeneutics is glossocentric, and it charges linguists, literary critics, and other bookish types with the task of explaining everything.

This doctrine was an offshoot of Dilthey's, and it was independently reinvented by Ludwig Wittgenstein and his followers, who tried to reduce all problems to "language games"; by Alfred Korzybski and the "general semanticists" who used to gather around the defunct magazine ETC; by Claude Lévi-Strauss and other "general semioticians," like the famous novelist Umberto Eco; by the existentialist Jacques Derrida, of "deconstruction" fame; and to some extent also by John L. Austin and his follower John Searle, who exaggerated the importance of "speech acts," like the judge's sentence "I condemn you to life prison." If you dislike manual work, try to "do things with words." This won't supply bread, but it may land someone an academic job.

None of these textualists engaged in scientific research or even paid any attention to the outstanding scientific findings of their time. Following Husserl's advice, they closed their eyes to the real world, and produced only texts or comments on them — whence the adjective 'textualist' fits them too. Some of their texts are just pieces of learned nonsense, like Heidegger's "The essence of being is IT itself," or silly wordplays, like Derrida's with the words *écrit* (text), *écran* (screen), and *écrin* (casket).

Computationism, whose slogan is "Everything is computation," is close to textualism in that it, too, attempts to reduce all facts to symbols. However, in contrast to textualism, computationism has produced some useful tools for handling the algorithmic, "mechanical" or rule-following aspects of intellectual work, in particular computer simulations of real processes, some of which are used in technology and in managing. Still, computationists skirt the most important real facts, or else offer us sanitized representions of them. In particular, information-processing psychologists explain nothing and do not help treat mental illnesses, because they ignore the organ of the mind.

8.6 Confusion Between Fact and Phenomenon

Historically, epistemological phenomenalism (*cognoscere est percipi*) preceded ontological phenomenalism (*esse est percipi aut percipere*). Moreover, one may endorse the former but not the latter, as Ptolemy did when he enjoined his fellow astronomers to "save the phenomena" (see Duhem 1908).

The concepts of fact and phenomenon are often conflated although phenomena, such as perceptions, occur only in nervous systems.

In effect, we know that nature is colorless, tasteless, and does not smell. But of course we also know that we see colored things when they reflect light exciting our visual system, and smell things that release molecules that are detected by nerve ends in our noses. In short, whereas phenomena occur only in brains, facts occur everywhere in the universe. Shorter: *Phenomena ⊂ Facts.*

The great Ptolemy rejected the heliocentric model of the solar system because he was an empiricist, whereas the realists Copernicus and Galileo adopted it. Yet there still are philosophers who do not know or do not admit the differences between facts in themselves, or noumena, and facts for us, or phenomena. For instance, Bas van Fraassen (1980) has claimed that the quantum theory allows one to calculate atomic phenomena. But of course there is no such thing as an atomic phenomenon. The smallest sensory detector is the feature detector in the mammalian visual system. The warning "for your eyes only" cannot be obeyed, because what can see is not the eye but the visual cortex when acted on by eyes stimulated by light pulses.

Other scholars defend or attack objectivism, or epistemological realism wihout concern for its ontological support. Max Weber (1904) devoted it his most brilliant and widely read methodological essay, not realizing that it contradicted Wilhelm Dilthey's philosophy, which he expounded and endorsed in his magnum opus (Weber 1976 [1922]).

One century later the historians of science Daston and Galison (2007) published a voluminous and lavishly illustrated volume on aesthetic objectivism, or the realism of figures such as anatomical drawings, maps, and portraits. Philosophers and scientists are far less interested in the fantastic inventions of Hyeronimus Bosch and Maurits Escher, or in the distortions of Picasso and Dalí, than in the questions whether social scientists can prevent partisanship from interfering with objectivity, and whether physicists can produce subject-free models of the physical world. Let us therefore tackle the perplexities that one of the greatest philosophers felt when thinking of the subject–object problem.

8.7 Kant's Indecision

The great Immanuel Kant was inconsistent on the realism issue. In fact, in the preface to his first *Critique* (1787b: 33) he held that it was "a scandal for philosophy and human reason in general" that some philosophers had denied the reality of the world. But in the heart of the same work (Kant 1787b: 724) he aserted that "*die Welt ist eine Summe von Erscheinungen,*" that is, the world is a sum of appearances. A few pages later he warned that the question whether there exists something that is not an object of possible experiences is "meaningless." Clearly, when he wrote this he had forgotten his earlier statement, that the denial of the objective reality of the world was scandalous.

Kant's successors, the neo-Kantians and positivists, from Comte to the Vienna Circle, embraced both Kant's ontological phenomenalism and his thesis that the problem of the existence of things-in-themselves is meaningless. In particular, the polymath John Stuart Mill (1843) and Paul Natorp (1910), the leader of the Marburg school of neo-Kantians, defined "thing" as "a set of possible experiences." They should have concluded the nonexistence of places, such as the young Kant's nebulae (galaxies), devoid of beings capable of having experiences.

For a while Alfred N. Whitehead and Bertrand Russell embraced phenomenalism, though without offering arguments. Eventually Russell (1940) changed opinion, adopted scientific realism, and proposed some interesting examples. One of these, translated into my lingo, is that every photon is generated by the transition of an atom from one energy level to a lower level. Physicists can measure the emitted energy but not the energies of the levels in question: these can only be calculated — as well as altered in a predictable fashion by the application of magnetic and electrical fields.

8.8 Refutations of Irrealism

When the famous wordsmith Samuel Johnson learned that George Berkeley had denied the autonomous existence of the world, he kicked a stone, to show his disagreement. But resorting to praxis is a didactic prop rather than a philosophical argument.

Quine and Goodman (1940) invented the cheapest way to undo things with words, namely to "eliminate the extralogical predicates" of a theory by tranforming its postulates into definitions, while suppressing the hidden existence assumptions, such as "the reference class of predicate P is nonempty" (Bunge 1967b, vol. 1: 132). Whoever adopts the said conventionalist trick should also propose closing down all laboratories and observatories. Needless to say, no one has used the Quine–Goodman trick, and eventually Quine adopted physicalism (personal communicatiion c. 1970), while Goodman became a constructivist — the most fashionable version of idealism.

Most Christian theologians are realists. They are satisfied with the tacit thesis of the book of Genesis, that being and the sacred are the same. It would be plasphemous for a Christian to deny the identity "Universe = Creation."

The most popular modern argument in favor of realism is the adequacy of the accepted scientific theories to their referents. But such adequacy only suggests that those theories are not sheer fantasies but more or less adequate (true) representations of their objects. The truthful narration or faithful pictorial represention of a nightmare, such as Goya's about the sleep of reason, are not about reality but about the dreamer. The mistake incurred in accepting the argument in question came from confusing the referents involved in it.

Since truths about the world do not prove its reality, let us try the dual strategy: let us see where falsity may lead us, as I did in my "New dialogues between Hylas and Philonous" (Bunge 1954). Gilbert Ryle, the editor of *Mind* at the time, professed to like it, but could not violate the journal's policy of rejecting dialogues. Lucky Plato, Galileo, and Berkeley!

Let us next consider the membership of the Flat Earth Society. What do we do when informing them that the Magellan circumnavigation of our planet falsified their dogma half a millennnium ago? What we do is to confront their belief with a datum concerning reality, and we give preference to the latter. That is, we confirm the reality of a feature of a part of the world by showing that at least one representation of it is false. And we close with an ad hoc new proverb: *Esse est errare.*

8.9 Scientific Research Presupposes Realism

There is no need to try and prove the reality of the external world, for we presuppose it in everyday life. Suffice it to recall what we do every time we wake up: we become aware of our immediate surroundings and start navigating in it, avoiding the obstacles in our way. That is, we behave like realists even while professing some irrealist fantasy.

True, a few eminent explorers, in particular quantum physicists, have claimed that the world is a creation of theirs. But they have not bothered to exhibit any evidence for this extravagance. If one analyzes the basic ideas of the theories they invoke, one fails to find any reference to observers or experimenters (e.g., Bunge 1967a; 1967b; 2012a).

For example, the hamiltonian or energy operator of a free "particle," such as an electron in outer space, contains only one term, namely the operator representing its kinetic energy; if the object of study or referent is a hydrogen atom, one adds a term representing the potential energy of the electron in the proton's electric field; and if the atom is embedded in an external electric or magnetic field, one adds a term representing its potential energy in the field. One would look in vain for terms representing an observer or even a measuring instrument: in all of the above cases only natural physical entities (particles and fields) occur. The references to observers and pieces of apparatus occur only in purely verbal comments that are so many arbitrary philosophical grafts. One might as well add "God willing."

8.10 From Herodotus to Quantics

Sometimes Herodotus, the putative father of European historiography, fantasized out of patriotic zeal. For example, he held that the huge Persian cavalry watered in a minuscule Attic stream. By contrast, we trust Thucydides's *History of the Peloponnesian War* because he fought on the losing side, did not magnify the victories of the Athenians, nor flattered their improvised generals. So far as we know, Thucydides stuck to the facts as provided him by his informants, all of them eyewitnesses like himself. In short, Thucydides was more truthful than Herodotus.

Two millennia after having swallowed the hagiographies, nationalist chronicles, lies, and half-truths of spiritualist historians, we honor Leopold von Ranke, calling him "the founder of scientific historiography," even though this title belongs to Thucydides. We honor Ranke because of his insistence, in the midst of Romantic excesses, that the historian's task is to tell "*wie es wirklich war*," that is, what really happened, instead of suppressing, exaggerarating, or lying to favor homeland or ideology, as it still happens. A recent example of blatant historiographic lie is the story that the British-American forces, not the Red Army, finished World War II by destroying the German army and the last Berlin bunkers.

In other words, Ranke, just like the geologists and chemists of his time, expected his colleagues to adopt the definition

To tell truthfully fact F = to tell how F really happened.

Briefly, for Ranke, as for all the scientific realists, every description of a real fact *F* ought to be congruent with *F*. We may also write: *If proposition p states the occurrence of fact F, then p is true if and only if F is real.* Thus "true" and "real," though not cointensive, are coextensive.

The subjectivists, in particular the constructivists, have no use for the concepts of factual truth and falsity, because they stipulate that the world is such as anyone may wish it to be: for them "anything goes," as Paul Feyerabend famously said when he exaggerated the skepticism of his erstwhile teacher Karl Popper (see Stove 1982). By contrast, to a realist the concept of factual truth, or adequacy of idea to fact, is central.

However, we have no reliable data about the role that truth plays outside science and technology. It would be particularly interesting to know the role that reality denial played in the most egregious failures in recent political and business fiascos, such as the Great Depression, the ascent of Nazism, the collapse of communism, and the war on Iraq. And how effective would such such studies on idea/fact mismatches be?

Although the idea of adequacy of idea to fact is central to science, we must admit that the sartorial notion of adequacy or fit, which works well at the tailor's, where two material and perceptible things are compared, namely bodies and clothes, fails when comparing, say, the formula of a chemical reaction, which is a conceptual object, with the reaction itself, which is a material one. In this case we must exactify the intuitive concept of truth of fact, an operation that requires some mathematical concepts (see Bunge 2012b). Suffice it to note that, whereas Leibniz (1703) noted the contrast between *vérités de fait* and *vérités de raison*, Tarski (1944), called "the man who defined truth," conflated them. Cases like this confirm the popular belief that there is no progress in philosophy.

8.11 Practical Philosophy: Six Realisms

So far we have been dealing with the realisms and their duals belonging in theoretical philosophy: logic, ontology, epistemology, and methodology. In the following we shall examine briefly the axiological, praxiological, ethical, aesthetic, legal, and politological reality concepts belonging in practical philosophy. However, we shall confine ourselves to defining them with the help of a dictionary (Bunge 2003a).

Radical axiological realism, characteristic of Plato, holds that all values are objective ideas and, moreover, precede valuable objects as well as their evaluators. In contrast, **moderate axiological realism** maintains that, whereas some values are objective and found, others are subjective and invented; that in the external world there are not values in themselves but only worthy objects; that everything valuable is so only in some respects; and that only organisms are capable of evaluating, so that values emerged nearly 3,000 million years along with life (Bunge 1962a; 1985b). These assumptions imply that human evaluations have changed not only along biological evolution but also along human history. Suffice it to remember the recent devaluation of asceticism and physical bravery.

Praxiological realism shares the general traits of realism but is confined to deliberate human actions. **Ethical realism**, a part of the former, denies the reality of Good and Evil in themselves but it holds that we may and even must rate — as good, bad, or indifferent — the intentions and actions that may affect other living beings either directly or through their environments.

Contrary to popular belief, not all valuations are subjective; sometimes we can and must find empirical evidence for or against such statements as "mutual aid is good," "unprovoked aggression is bad," "pronounced social inequality sickens," "ignorance is dangerous," or "the state is the custodian of the commonwealth" — whence "libertarianism is just as antisocial as tyranny."

Finally, **aesthetic realism** holds that beauty and uglinesss are objective and universal. The moderate aesthetic realists will confine themselves to noting that the artistic values are made and unmade in the course of history, and that some of them are embraced by all the members of a human group: *ars filia temporis et societatis*. For instance, among educated people in Western countries nowadays almost no one likes kitsch, "socialist realism," metallic rock, or the theatre of the absurd.

On the other hand, among the same people the Parthenon and the Pompeii mosaics, as well as the Alhambra and Notre Dame, Beethoven's symphonies, and the impressionist paintings are still being admired. In short, in matters of art, subjectivism works for some whereas realism works for others, and garbage for still others. Art, then, seems to be *the* sector of culture where relativism rules.

On a very different note, **legal realism** invites us to look at law in the making rather than at the received legal corpus (e.g., Lundsted 1956; Pound 1954; Stone 1966). To a legal realist, the legal code is a part of the manual of social behavior and political governance. That is, realists look at legal codes as the transient products of legal discussions and fights involving lawmakers, politicians, and legal activists as well as at attorneys and the sociologists and historians who study how and why laws are written, repaired, and broken. The current debates and fights over racial discrimination and the reproductive rights of women constitute a case in point, for the pertinent laws are being loudly challenged in public spaces, revised in parliaments and courts of law — and in Michael Moore's satirical films.

Although legal realism looks like the most realistic legal philosophy, it is far less popular than its rivals, legal positivism and natural law. Legal positivism is nothing but a form of confomism, for it will approve of the death penalty wherever it is practiced, and reject it elsewhere Thus, although it claims that the law is morally neutral, this school is morally and politically abhorrent (Bruera 1945, Dyzenhaus 1997).

As for natural law, this very expression is absurd, since a look at the history of jurisprudence shows it to be a fragile artifact. Just recall the Roman law about the use and abuse of all private goods, including slaves. Clearly, the rule of law changes along with society: what used to be regarded as a just rule may now be regarded as unfair and vice versa. Moreover, litigants, their lawyers, and judges will now use legal categories, such as that of a war crime, as well as admit pieces of scientifc evidence, such as DNA sequencing, unthinkable until recently. In sum, the very concept of natural law is an oxymoron. Still, the natural law school contains an important grain of truth absent from legal positivism, namely the idea that any law can be morally justified or challenged: the legal and the moral intersect, so that Just ≠ Legal.

Finally, *political realism* comes in two versions: scientific and ideological. The former includes moral precepts along with practical rules, and it is confined to noting that (a) politics has two sides, the struggle for power and its exercise, or governance; (b) no human group can escape politics;

and (c) political scientists pay more attention to political dealings and conflicts than to the rhetoric, often mendacious, that mobilizes or paralizes people, so that, unlike the postmodern writers, they should not confuse political action with political narrative.

For example, realists will interpret the Crusades and the Discovery Voyages as business ventures rather than as missions to save heathens's souls. Likewise, the current Islamic bellicosity may be seen as a primitive delayed reaction against the Western aggressions on resources-rich nations (see Fontana 2011). This cynical way of understanding political realism uncovers the real aim of the so-called war on terrorism, since war is terrorism at its worst. Intelligent realists seek peace through negotiation and cooperation, not war, as the European statesmen finally realized when they signed the Westphalia Treaty of 1648, which ended the Thirty Years' War.

By contrast, the authors of the Treaty of Versailles (1919), which ended World War I, trampled on the principle of self-determination of the peoples, on top of which it imposed on Germany conditions onerous and humiliating enough to breed the Nazi reaction — as Keynes foresaw. Do we really learn from our mistakes?

8.12 The Antimetaphysical Reaction

Metaphysics, or ontology, was the nucleus of philosophy until the modern age. When science was reborn in the heads and hands of Copernicus, Kepler, Galileo, Descartes, Vesalius, Boyle, Harvey, Huygens, Torricelli, and a few others, metaphysics was quickly discredited in the West: it was seen as part of a backward worldview and a servant to the ruling religion. In other words, the indifference or hostility of the schoolmen to the unexpected novelties of the Scientific Revolution begat the idea that all metaphysics are dispensable or even despicable.

It was thought that, since the new knowledge did not derive from reading old books but in spite of them and often against them, it called for a new theory of knowledge, which had to be built to facilitate further advances. Only Hobbes and Spinoza were shipwrecks of metaphysics.

The critics of scholasticism, from the medieval nominalists to Francis Bacon, tried to divorce epistemology from ontology. The phenomenalists went even further: they attempted to reduce ontology to their own subjectivist epistemologies. George Berkeley (1710), the most original, radical, eloquent, and outrageous of them, found holes in the budding infinitesimal calculus, but did not dare assaulting Newton's mechanics (1687), the first successful scientific theory in history. By contrast, David Hume (1748) had the nerve to criticize it although he did not have the mathematics required to understand it.

Whereas Berkeley and Hume realized that they were swimming against the current, Kant paid lip service to Newton, and believed to have performed a "Copernican revolution" in philosophy, while in fact his subjectivism was anti-Copernican. However, Kant was not honored during his lifetime. In particular, the French Enlightenment ignored Kant.

Kant was severely criticized by a few contemporaries. In particular, the eccentric genius Johann Heinrich Lambert — a friend of Leonhard Euler and Daniel Bernoulli — tried in vain to persuade Kant that he was wrong in conceiving of space and time as subjective. In his letter of October 13th, 1770, Lambert informed Kant that change is linked to time, and cannot be understood without it. "*If changes are real so is time, whatever it may be. If time is not real, then no change is real*" (apud Kant 1912, vol. 1, p. 101, original emphasis). I have seen no evidence that Kant replied. His only reasonable reply would have been to give up his subject-centered metaphysics.

Yet Kant's philosophy triumphed in academia in the wake of Hegel's attack on the Enlightenment. In fact, during the whole 19th century, academic philosophy was ruled by Kant and Hegel, at the same time that the sciences of matter were being consolidated. Toward the end of that period a few distinguished scientists attempted to expel matter from science. Let us peek at them.

8.13 Dematerializing the Sciences of Matter

The concept of matter was the *bête noire* of idealism since Berkeley's *Treatise* (1710). Scientists rarely use the word because it belongs to the philosophical lexicon, but they cannot help using its logical subordinates, *body*, *reactant*, *organism*, *individual*, and *merchandise* — none of which could be interpreted in spiritual terms without severe distortions.

Wilhelm Dilthey called the social sciences *geistig* (spiritual), but none of his followers claimed that its referents, namely real people, were immaterial beings who could live without eating and excreting material things. But they did claim that everything social is spiritual or cultural — a reduction that no war veteran will appreciate.

How might the concept of matter be expelled from science without resorting to tricks? Ernst Mach (1893), an outstanding experimental physicist and psychologist, thought he had found the way. His alleged proof can be analyzed into three steps. First, define 'material' as "massive," even though the electromagnetic fields — ignored in Mach's *History of Optics* — are massless. Second, consider a system of two bodies attached by a spring. According to Newton's second law of motion, the force that keeps them together is $F = m_1 a_1 = -m_2 a_2$; (c) consequently the mass of body 1 relative to that of body 2 equals $m_1/m_2 = -a_2/a_1$. Third, lo and behold: the guilty concept of relative mass has been reduced to the innocent one of relative acceleration.

Mach should have been told that his alleged proof was circular, for his alleged proof involves the very concept he intended to eliminate. Nevertheless, countless textbooks have repeated Mach's fallacy over one century — a sad comment on the logic of college physics teachers (see Bunge 1966; 1968). In short, Mach failed to refute materialism. Moreover, his discovery of the bands named after him constituted a valuable contribution to cognitive neuroscience, a child of materialism, since those bands are caused by the inhibition of the neurons in the retina that flank those that are excited by the central luminous stimulus. Welcome to the materialist camp, Herr Professor Mach!

Energetism was another weapon of the anti-materialist assault against materialism around 1900. Its gist was the thesis that the universe was made of energy, not matter. Its champion was Wilhelm Ostwald, an eminent physical chemist and Nobel laureate. In his 1895 allocution he claimed that energetism had overcome scientific materialism. And, just like Mach some years earlier, he also attacked the atomic theory — until 1908, when he admitted that experiment supported it.

Anyone who has read a formula like $E(a,f,u) = n$ for the energy of thing a relative to reference frame f, and in unit u, knows that E is a not an entity but a *property* of some concrete (material) entity *a*. For example, the famous relativistic formula "$E = mc^2$" holds for any thing endowed with a mass *m*, whether it be body or particle. Its counterpart for photons, which are massles, is "$E = h\nu$," where *h* is Planck's constant — the badge of everything quantic — and ν designates light frequency. In short, every energy is the energy *of* some concrete entity, hence not a candidate for a basic constituent of the world. Consequently, energetism was not viable. Yet, a century ago it enjoyed the support of some thinkers, who believed it to overcome both materialism and spiritualism.

8.14 Joining Metaphyics with Epistemology

In principle, metaphysics and epistemology are mutually irreducible, for while the former is about all entities, epistemology is confined to our study of them. For example, it is logically possible to admit at the same time the materialist idea that ideas are brain processes, and that the brain invents the world. Likewise, it is logically admissible to hold that the world exists by itself, and that ideas reside in the immmaterial mind. In short, there are four logically possible combinations of materialism *M* with realism *R* and their negations. I have proposed the *MR* combination and argued against the other three (e.g., Bunge 2006; Mahner 2001).

Though distinct and logically separate, ontology and epistemology are *de facto* interdependent. The reason is that, while the former tells us what kinds of objects can exist, epistemology examines the reasons we have to assert that they do or do not exist. However, this examination cannot even be planned unless we assume something, if only as a working hypohesis, about the nature of the object of knowledge: that it is real or imaginary, physical or social, knowable or mysterious, and so on. This is why we ought to join epistemology with ontology. If we add compatibility with the bulk of contemporary science, we get what I call *hylorealism,* that is, the fusion of realism with materialism (Bunge 2006). As a matter of fact, all philosophers have employed hylorealist categories, such as those of appearance, thing in itself, and worldview.

Hylorealism frees us from phenomena without things in themselves (Berkeley, Hume, Kant), multiple possible universes (Putnam, Kripke,

David Lewis), knowledge without knowers (Popper), brains in vats (Putnam), zombies (Kripke), and similar idle fantasies. An additional advantage of hylo-realism is that it is a research program — hence an unemployment insurance.

Finallly, how real is virtual reality? Virtual reality technology designs artifacts that produce illusions, such as those of flying or of seeing the world upside down. Obviously, all such simulations are real, but none of them give us objective views of the world out there. Some of the virtual reality gadgets are used for amusement, others for instruction, and still others to find out selected aspects of animal behavior, such as the effect of visual stimuli on a rat's navigating or exploring its environment. In short, virtual reality differs from external reality in that it involves a perceiving animal and is crafted with the help of electronics and computer science. It might be better be called *simulated perception*. The philosophical lesson is of course that *Reality ≠ Objectivity*.

Coda

Philosophical realism was taken for granted in the West till Galileo's trial in 1633. Since then, most philosophers have opposed realism at the same time that scientists and technologists practiced it.

The first scientists to deny the independent reality of reality were the quantum physicists who, in some of their popular publications (never in their technical papers), claimed that quantics had confuted realism. Einstein complicated the controversy by confusing realism with classicism — the dogma that all things must possess the properties predicated by classical physics (Bunge 1979a).

Half a century later a second confusion arose, namely that involving so-called *local realism*, related to entanglement, or loss of individuality. The real intent of the statement that "local realism has been experimentally confuted" is to state that entanglement involves the downfall of the principle of *action by contact*. According to this principle, two mutually separated things can interact only through a third one, such as a field, interposed between them. The loss of this principle, entrenched in field theories, has nothing to do with realism, and everything to do with the systemic motto "Once a system, always a system." Newton would be puzzled but Leibniz might be delighted, for he upheld the unity and continuity of the world.

In my view, the debate over the reality of a theory is settled by axiomatizing it, since this operation boils down to finding and analyzing the theory's basic concepts, and checking whether any of them refers to observers. I was thus able to *prove* that nonrelativistic quantum mechanics is just as realistic as all the other basic physical theories (Bunge 1967a).

My long-distance students Guillermo Covarrubias and Héctor Vucetich, just like the latter's students Santiago Pérez Bergliaffa and Gustavo Romero, corroborated and refined my realist reconstruction of quantum mechanics and general relativity. None of these research projects challenged the counter-intuitive components of the theories in question, in particular the coherence and decoherence (or projection) of quantum states, the entanglement of the components of a quantum system, or the jelly-like nature of spacetime. If these features were not real, they could not have been experimentally confirmed with amazing precision (Zeilinger 2010).

In sum, scientific realism has been vindicated, whereas irrealism has fallen by the wayside, if only because it has not inspired a single successful research project. I'll tell you what your philosophy is worth if you tell me what research it has inspired (Bunge 2012a).

CHAPTER 9

MATERIALISMS: FROM MECHANISM TO SYSTEMISM

Materialism is the set of philosophies according to which all the constituents of the universe are material rather than spiritual. Although materialism is sometimes confused with realism, the two are logically independent from one another, since materialism is an ontology, whereas realism is an epistemology. Consequently, the four possible combinations <M,R>, <not-M,R>, <M,not-R>, and <not-M,not-R> are logically possible.

Materialism was tolerated in ancient India until the Moghul invasion, but it has had a bad press in the West since Plato. It has been particularly vilified since Justinian imposed Christianity as the state religion and incited mobs of fanatics to destroy all the remains of pagan culture. The main reason for this intolerance for materialism is that it entails atheism, hence secularism. But the ban on materialism could not prevent it from continuing to be the spontaneous if silent ontology of scientists and physicians.

9.1 From Early Materialisms to the Scientific Revolution

Some of the best-known materialist philosophers in Western ancient philosophy were Democritus, Epicurus, and Lucretius. Before the Christian attack on secularism, the Epicureans had been just as popular as the Christians, Stoics, and Mithraists. After Justinian imposed Christianity as the state religion, and decreed that the emperor was also the pope, Epicureanism all but disappeared along with the other non-Christian sects. From then on, materialism survived during one millennium in India (Charvaka and Samkhya) and in the Islamic countries (Averroes), followed in Western Europe by the Latin Averroists.

In Western Europe, materialism resurrected around 1500 along with science, and mainly among the Aristotelians interested in natural science and the philosophy of mind. Many of these materialists and cryptomaterialists called themselves Aristotelians and, in fact, so they were. Indeed, Aristotle's mechanics and marine biology had no use for immaterial entities, and some Aristotelian philosophers attacked the dogma of the immortal soul by sticking to The Philosopher's definition of the soul as "the form of the body," which implied that it vanished upon death. This was a central theme of Cesare Cremonini's, the popular colleague and rival of Galileo's at the great University of Padua (see Renan 1949).

Paradoxically, the Aristotelian philosophers did not hail the Scientific Revolution. Instead, they kept defending and commenting on their ancient mentor's theses, whereas the core of the new science was a whole sheaf of original research projects, such as Galileo's research on moving bodies, Gilbert's on terrestrial magnetism, Pascal's on the relation between atmospheric pressure and height, Vesalius's on human anatomy, Boyle's on chemistry, and Harvey's on the circulation of the blood. All of them took it for granted that nature was material, but their philosophical colleagues hardly noticed the Scientific Revolution.

9.2 Descartes, the Anomalous Philosopher–Scientist

Descartes, usually dubbed "the father of modern philosophy," was also called "the masked philosopher" for, although he professed to believe in God and in the immaterial and immortal soul, he also conjectured that this was the function of the pineal gland, which is a brain organ. In short, Descartes was a psychoneural dualist in some of his works, and a monist in others.

The Catholic hierarchy was not fooled: it placed all of Descartes's works in the *Index Librorum Prohibitorum*. The materialist philosophers of the French Enlightenment were not fooled either: they regarded Descartes's treatise on the world (1664) as cryptomaterialist. And the man himself, outwardly a devout Catholic, escaped the clutches of the Inquisition: he first fled to Calvinist Netherlands, and finally to Lutheran Sweden.

Briefly, when Modernity came, materialism remained ensconced in the newly born scientific community, whereas most philosophers remained in the spiritualist camp. Descartes managed to step one foot on each camp, for he split the set of entities into extended or material, and thinking or immaterial.

The official story of modern materialism is that it is a sequence of a few high but inconsequential peaks, such as Spinoza, Gassendi, and Hobbes. All three were near coevals, soon to be superseded by the far more sophisticated and influential Berkeley and his successors — Kant, Hegel, and their intellectual progeny: Comte, the Marburg neo-Kantians, Mach, Dilthey, Husserl, Bergson, Croce, Gentile, and present day's postmodernists, especially the social constructivists.

Let us focus for a moment on the Scottish branch of the Enlightenment, which included the two Davids, Smith, and Hume. The one who became the most popular philosopher of all times according to a recent opinion poll, was David Hume. He is better known for his overall skepticism than for his own version of empiricism (1739), which owes much to Berkeley's. But, contrary to Berkeley, Hume was a philosophical naturalist, since he was non-religious, and held reason to be the slave of passion — that is, he took knowledge to be but a survival tool — though he did not care to elaborate. In any case, his philosophy was irrealist, immaterialist, and politically conservative — just the opposite of the radical fringe of the French Enlightenment.

9.3 Materialism Among the Philosophers

Friedrich Lange's popular *History of Materialism* (1866) was a much more detailed treatment of materialism than any of the general histories of philosophy — those of Harald Høffding, Emile Bréhier, and Bertrand Russell. But Lange, a follower of Nietzsche, offered no analysis of the scientific theories of his time in support of his claim that they supported idealism (Lange 1875 vol. 2: 533–534). Moreover, he claimed that, what Kant had called "things in themselves," are actually things for us, since we use the same ideas to describe both. The Kant experts are still debating this point.

Most of the philosophy courses around the world overlook the materialists and cryptomaterialists, particularly those belonging to the radical fringe of the French Enlightenment — Holbach, Diderot, Helvétius, La Mettrie, Maréchal, Cloots, and their friends. At McGill University, where I taught philosophy for half a century, I was the only one to deal with those undesirables, whereas two colleagues offered courses on Noam Chomsky's scattered philosophical opinions.

The cases of positivism and dialectical materialism are more complicated: Comte's positivism resulted from combining Kantianism with scientism, and dialectical materialism from combining Hegel's murky dialectics with French materialism. The logical positivists claimed to have superseded the materialism/idealism chasm by ignoring it, and the Marxists thought to have refined vulgar or mechanistic materialism by taking in Hegel's absurd dialectics along with Feuerbach's simplistic materialism, far inferior to the one Holbach had crafted one century earlier.

The result of the latter synthesis, informally called *diamat*, was the official philosophy of the Communist Party, and as such immune to critical analysis — which explains its ossification despite its many confusions and mistakes. For example, Lenin defined 'material' as "having the property of belonging in objective reality," or "existing outside our consciousness." This confusion of materiality with objectivity may have led Lenin to supporting psychoneural dualism, which in turn explains the backwardness of psychology in the former Soviet Union, where psychoneural dualism was the party line, and only lip service was paid to Pavlov's crude materialist schema.

9.4 Max Weber, Antimaterialist but Realist by Half

It ought to be obvious that those who study reality must adopt philosophical realism. This was clearly understood by Emile Durkheim (1895), one of the founders of modern sociology, who insisted that social facts should be regarded as being just as real as physical facts. Max Weber, the other putative father of the discipline, was not blessed by the Cartesian clarity that the pre-postmoderns used to brag about. In fact, Weber oscillated between objectivism (Weber 1904) and Dilthey's idealistic hermeneutics (Weber 1921). Let us take a quick look at his work for, unlike Durkheim's, it has exerted a strong and lasting influence on the philosophy of the social. This influence was largely due to the support he got from the neo-Kantians, and above all because he was seen as the anti-Marx — which he was not, for he lacked Marx's originality, social sensibility, and political courage.

Weber is best known for his idealist thesis that Calvinism begat capitalism because both favored austerity and saving. But of course everyone knows that before saving money one must make it, and that merchant capitalism was born in northern Italy (the Medicis) and southern Germany (the Fuggers), not in small and austere Geneva, Calvin's birthplace. It is also well known that the cradle of industrial capitalism was Manchester, not Amsterdam — the other seat of Calvinism; and that none of those riches would have been obtained without the profits from mining and the trades in spices, precious metals, mercury, silk, opium, and above all African slaves sold to work in the American and Caribbean plantations.

In short, Weber's idealist thesis about the capitalism–Protestantism connection is utterly false. So is his fantasy that the Indian caste system came ready made from the head of a smart Brahmin scholar, not as a result of a long history of invasions and conquests. Neither of these opinions of Weber's was supported by empirical evidence.

However, fortunately for us, Weber was not a consistent idealist. Indeed, the very same year of 1904 that saw his most famous albeit weakest publication also saw his eloquent defense of objectivism, or realism, in social science. But this brilliant paper was nullified by his praise of Dilthey's *Verstehen* (understanding) method. According to Dilthey, what the social scientist (*Geisteswissenschaftler*) must do is to capture the intentions of the actors — surely a matter of free guessing.

Ironically, his own *Verstehen* did not help Weber understand the real intentions of the political and military leaders of the Great War, whose continuation he favored till the bitter end. Nor was this his only blind spot. Indeed, most of his writings are about the history of dead ideas, not the great events that fashioned his time. In fact, Weber paid no attention to the Industrial Revolution; to the colonial adventures on all continents; to the investment of the profits of the slave trade into the English industry; to the emergence of science-based industry; to the rise of cooperativism, trade-unionism, socialism, feminism, and secularism — or to the extension of the scientific method to the social and biosocial sciences and technologies.

In sum Weber, unlike Marx, did not understand his own times and had no impact whatsoever on his country's fate. This failure may have resulted from his bookishness combined with his attempt to match his idealist metaphysics to his two mutually incompatible epistemologies, the realism inherent in science and the subjectivism of the *Verstehen* school.

9.5 The Reception of Materialism by Modern Scientists

The official story of philosophy overlooks the crucial fact that, while shunned, distorted, or vilified by most professional philosophers, materialism triumphed sensationally in all the natural sciences from Galileo onwards. Indeed, all the modern natural scientists have studied nothing but material things on the physical, chemical, and biological levels. Only spoon benders and psychoanalysts still write about the power of mind over matter.

It might be argued that concrete things dwell in spacetime, which would be immaterial. However, the central equation of Einstein's theory of gravitation, aka general relativity (GR), tells us a far more complex story, namely that spacetime would vanish along with matter. Indeed, the equation in question can be condensed into "$G = T$," where G describes spacetime while T, the so-called matter–energy tensor, describes the sources of the gravitational field.

If the matter tensor T were to vanish everywhere, that is, if the universe were hollowed out, nothing would remain, not even empty space. Hence, so far as GR is concerned, real existence and spatio-temporality are coextensive: neither exists without the other. In other words, spacetime involves matter and conversely, so that what is real is *matter-in-spacetime.*

Separating spacetime from matter is just as wrong as detaching motion from moving things, as Aristotle would say. Thus, physicists use the concept of matter even if they do not employ the corresponding word. And the universe "wears" only one of the geometries, namely Riemann's. All the other geometries are imaginary, although Euclid's is an excellent approximation for mid-size bodies and quantons.

Should we take one more step and assert that spacetime is material? Let us start from the evidence for the reality of gravitational waves provided in 2015 by the LIGO team. Since such waves are but ripples in spacetime, it follows that the latter is closer to a jelly-like grid than to a rigid one. Now, in our ontology "x is real" is identical to "x is material," which in turn amounts to "x is changeable." Since the LIGO folks have shown that spacetime is changeable, they have unwittingly proved also that *spacetime is material.*

Admittedly, this conclusion is puzzling, hence deserving of a detailed physico-philosophical research project that would overflow the present work. Suffice it to add that, despite the many dematerialization efforts made not only by idealist philosophers from Berkeley on, but also by some outstanding physicists from Mach to Wheeler, the concept of matter is firmly entrenched in physics, which may be regarded as the most basic science of matter, hence as a vindication of materialism.

Materialism has also made some significant inroads in the social sciences, particularly historiography. In fact, like Ibn Khaldûn five centuries earlier, the members of the *Annales* school, which flourished around 1950, always started off by finding out where ordinary people lived, what they did for a living, what they ate, and how they behaved to one another: they practiced the ancient injunction *Primum vivere, deinde philosophari* (see, e.g., Braudel 1982, Schöttler 2015).

Materialism has also been very influential on the emerging biosocial sciences: anthropology, archaeology, social geography, demography, and later on psychology, epidemiology, and education science. For instance, nobody can deny that poverty, in particular malnutrition, delays child development, which in turn explains poor scholastic achievement (Cravioto 1958).

In short, the greatest achievement of materialism in modern times is not its breeding atheism, which had preceded science (for instance in original Buddhism), but its having inspired atomic physics and chemistry, "mechanistic" (non-vitalistic) biology, evolutionary biology, scientific anthropology and historiography, and cognitive neuroscience — whose guiding principle is "Everything mental is cerebral." Incidentally, this principle shows that materialists do not underrate ideas; they just place them where they happen.

Philosophy too has been influenced by materialism, as shown by the recent history of ontology, epistemology, and ethics. A naturalistic ontology is couched exclusively in physical terms, as was the author's theory of spacetime in terms of changing things (Bunge 1977). I have also done my bit to expel spiritualism from physics and chemistry (e.g., Bunge 1967a; 2010), and to craft a metaphysical theory both materialist and compatible with contemporary science (Bunge 1959a; 1959b; 1977; 1979b; 1980).

Quine (1969) famously proposed to construct a "naturalized epistemology." Regrettably, he identified psychology with its behaviorist version, which denies the mental, and refused to acknowledge propositions (see Bunge 1975). My own contribution to this project (Bunge 1983a) has been to treat cognition as a process in a socially embedded brain, and psychology as a biosocial science (Bunge 1987). This characterization of the discipline conforms to both Hebb's hypothesis about new ideas as the combination of neuron assemblies, and to the classical experiments of the social psychologists showing the effects of the social surrounding on perception and behavior.

Naturalistic ethics may be compressed into the thesis that moral norms are natural because they favor natural selection. This moral philosophy has become rather popular in recent times despite the well-known fact that moral norms deal with social conduct, which is known to differ in different social groups. For instance, torture and the death penalty are abhorred by most of us, mostly viscerally, and by some because they know that legal cruelty does not deter crime.

In summary, the report card of philosophical naturalism is mixed: it was progressive at the time when organized religion was the main roadblock against social progress. But at present, when Wall Street and its political minions have assumed that role, naturalism has become politically irrelevant in the best of cases, and regressive in the worst.

At present, the social studies are the last redoubt of Max Weber's idealist half: there we still find scholars who replace 'society' with 'culture,' 'politics' with 'political discourse,' 'exploitation' with 'rationality,' and who claim that political events are the direct consequences of certain ideas.

Thus, in his richly documented book on the ideas fought over during the French Revolution, Jonathan Israel (2014) tried to show that the revolution was the fairly direct product of the progressive ideas and fiery speeches of a few dozen writers and orators. No doubt, such ideas helped orient the revolutionaries, but they might not have had so many followers in France and other countries if wealth had been better distributed, as in the less advanced societies, and better managed, as in Britain.

Social injustice is a far stronger political motivator than even the most eloquent political manifesto. The brilliant *Communist Manifesto* was hardly noticed when published in 1848; and surely it was not Mao's pathetic *Little Red Book* that a century later mobilized millions of illiterate peasants to fight at once and successfully their landlords and moneylenders as well as the Japanese invaders.

This is only one instance of the materialist conception of history, according to which material interests are more effective political motivators than lofty ideals. This thesis is true because the brain does not work well when ill fed. In other words, the liberty–equality–fraternity triad stands on the job–health–education tripod, not the other way round.

Thus, spiritualism is basically wrong, whereas dialectical materialism is wrong in the best of cases, and confused in the worst. In contrast, historical materialism has a narrow but solid core. This core is economism, a view that deserves being enriched with sociology, politology, and culturology, so as to constitute a total or systemic conception of society (see Bunge 1979b; 1998; Wang 2011). By contrast, the opinion that class struggle is the engine of history — the central thesis of historical materialism — fails to explain all the most salient historical events, from the rise of agriculture and the state to the Hun and Mongol invasions to colonialism and the two world wars.

9.6 Naturalism, a Precursor of Systemic Materialism

Naturalism is the worldview according to which everything real is natural, i.e., *Universe = Nature*. This ontology did much to undermine the religious cosmologies, in particular the dogma that the world was created by the omnipotent and omniscient pure spirit. Indeed, from the start of modernity, no scientific research project has assumed the existence of God or any other supernatural beings. Since the beginning of modernity, science has been fully secular.

Although Christians profess to love nature and regard it as God's creation, they have always treated it as man's servant, to be used and abused at will and without concern for its conservation. The respect and love of nature rose only at the time of Spinoza, and the worship of nature was part of Romanticism from Rousseau to Goethe to the naturists, nudists, sun-worshippers, free-lovers, and campers who flourished in Europe between ca.1870 and World War II.

In his *Émile,* read by millions during two centuries since its publication in 1762, Rousseau held that an intimate contact with nature was far more educational than formal schooling. The accompanying epistemological principle, that *sentiment* (feeling, emotion, passion) was superior to reason, was part and parcel of Romanticism, and it became more than an encumbrance to the cult of reason and the spread of scientism. For instance, the enlightened philosopher and politician Anacharsis Cloots was guillotined by order of Robespierre, a follower of Rousseau's version of irrationalism.

In short, Romanticism was revolutionary in literature, music, and the plastic arts, but reactionary in science and philosophy because it included the toxic combination of naturalism with irrationalism. This combination may have been a source of the political ambiguity of Romanticism, which included both left-wing radicals like Maximilien Robespierre — whom Lenin called "a Bolshevik *avant la lettre*" — and right-wing ones like Joseph de Maistre, a precursor of Catholic fascism.

The cult of nature all but disappeared along with the twilight of Romanticism and the spectacular success of the natural sciences, especially chemistry, around 1850. But naturism resurrected around 1880 as part of the social protest against the status quo. In fact, it often came associated with anarchism, and in Germany it was the ideology of an increasing number of youth organizations like the *Wandervögel*. This grassroots movement, initially devoted to organizing open-air group activities, was eventually hijacked by the *Hitlerjugend*, which gained the monopoly on summer camps. In these camps, young boys were taught military skills while girls were told that free love was all right as long as it bred future soldiers.

In recent years, naturism has resurfaced as the radical fringe of environmentalism, which condemns the cultivation of genetically modified organisms as well as "allopathy" (scientific medicine), regardless of scientific evidence.

The naturist's efforts to commune with nature have done nothing to protect it, much less to protect people from exploitation, colonialism, or military aggression. Only a systemic worldview shows that a sustainable environment comes together with a sustainable society. Thus, the social blindness of naturalism tacitly condoned not only social injustice but also an unsustainable exploitation of nature. This is the paradox of naturalism: that it can be used as a license to degrade what it worships.

9.7 The Supranatural Order

Naturalism is likely to get us into trouble outside natural science. In fact, it fails every time it encounters artifacts and social relations, for neither of them is natural. In fact, axiological and ethical naturalism are wrong, for not everything natural is good, whereas everything social, whether good or bad for the person or the community, is artifactual. In fact, we normally take precautions against natural disasters such as floods, forest fires, earthquakes, and plagues, and we welcome the artifacts that save us harmful or demeaning manual work.

Besides, since Hammurabi's Code (1728 B.C.), we also attempt to compensate for natural inequalities like the comparative weakness of children, women, the sick, and the old. Only important priests like St. Paul could afford to preach blind obedience to the weakest members of his unjust society — slaves and women.

In short, naturalism fails to account for *supra*natural entities or events, which should not be mistaken for *super*natural (or miraculous) objects. There is nothing mystic about supranatural beings, such as humans, machines, and social organizations. Indeed, humans are the fully unnatural animals: we educate and domesticate or enculture ourselves. Unlike all other animals, we design, make, and repair multifarious artifacts, from axes and huts to microscopes and computers, and from business concerns to schools and armies.

Animals of a few other species work too and, in particular, they make artifacts, from nests to beehives to dams, but, as Marx pointed out when comparing houses with honeycombs, we are the only ones capable of designing them. And, although all gregarious animals follow some rules of social behavior, we are the only ones who invent and alter them from place to place and from time to time, though not always to improve our chances of survival.

Indeed, humans are the only animals who sometimes engage in antisocial and even self-destructive behavior, such as exploitation, crime and "weak thinking," in addition to constructive behavior, such as education and research.

Naturalism ignores all that: it emphasizes that everything human is natural, while in truth much about us, from machines and hygienic rules to legal codes and science, is artificial. A few naturalists have achieved celebrity by stating that some of the worst social ills, such as corporate dominance, are natural outcomes of gene sorting by natural selection. This is of course the social-Darwinist view of the social crafted at the close of the 1800s by Spencer and a few other defenders of the powers that be — in particular the British empire, the largest in history, which thrived with the slave and opium trades.

A recent instance of this political companion of the naturalist ontology is the "general theory of evolution," which claims that everything, whether natural or social, results from natural evolution, and that we would be much better off if we let evolution operate, freeing individual resourcefulness and market forces from government regulations (Ridley 2016).

Needless to say, this "libertarian" version of naturalism did not result from any research projects; it is just old conservatism in new garb. Ironically naturalism, which was originally intended to free us from religious servitude, has been used to justify market servitude. Is there a more persuasive evidence for the thesis that we are artifactual as well as natural?

In short, biologism does not account for human society any more than physicalism does. So, we need to enlarge naturalism to include social matter. The result is emergentist or systemic materialism, to be examined anon.

9.8 Systemic or Total Materialism

Contemporary science suggests that we should admit that material entities, relations, and processes come in several mutually irreducible kinds: physical, chemical, biotic, social, and technological, and that some social items, in particular the cultural ones, are artifactual. For example, there was nothing natural about the "naturalistic" (realistic) cave paintings; and mathematical logic is thoroughly unnatural — as shown, for instance, by the principle of addition: "A entails A or B, where B need not be related to A."

The higher levels of organization have presumably emerged from the lower ones in the course of the history of the world. For example, the biotic level emerged from the chemical one a couple of billion years ago, but every living being has suprachemical properties, such as metabolism, the ability to capture environmental items, and to reproduce its kind.

To put it in negative terms, every one of the partial materialisms, in particular the better known ones — physicalism and biologism — fails to capture some levels of organization. To do justice to all the varieties of matter we need the most inclusive of all materialisms. We will call it *total* or *systemic* materialism.

Systemic materialism has at least two advantages over its rivals in the materialist family. One is *pluralism*, in the sense that it admits the multifarious qualitative variety of the furniture of the world as well as that of the disciplines that study it. For example, social science and biology are distinct though they overlap partially. The result of this combination, namely biosociological science, allows one to explain such facts as the apathy of poor children as a result of malnutrition and exclusion.

Incidentally, the mere existence of biosociological sciences destroys the wall between the natural and the cultural sciences erected by Kant, Dilthey, and other idealists. It also confutes the attempt to confute the myths that mental ability and social standing are inborn, whence schools are redundant, and an initial endowment is unnecessary to climb social ladders.

Another advantage of systemic or total materialism is that it defines a *general concept of matter*, namely as whatever is changeable — or, if preferred, whatever can be represented by a state space with more than one element (Bunge 1977). Compare this definition with Russell's (1954: 384): "A piece of matter is a logical structure composed of events." Hence, "Electrons and protons [...] are not the stuff of the physical world: they are elaborate logical structures composed of events" (op. cit.: 386).

How does this bizarre combination of things with their conceptual models differ from idealism? And how does it help scientists tell the technical artifacts occurring in experiments from natural properties? Surely not even radical idealists would claim that logical structures may be accelerated by electric fields and bent by magnetic ones.

Coda

We have argued that modern science endorses materialism. It might be objected, though, that immaterialism is still going strong in psychology, as shown by the popularity of functionalist or information-processing psychology, which asserts that matter does not matter in the study of the mental. However, an inspection of the current literature will show that the only enthusiastic functionalists are philosophers; the scientific psychologists share the idea that all mental events are brain events.

This is why, unlike their predecessors, today's psychologists study brain organs, like the amygdala, in particular their specific functions (properties and processes), just like physicists study moving bodies rather than motion separately from the things that move, as Plato recommended. The functionalists, eager to get rid of matter, have not heard that Aristotle was right in criticizing Plato's doctrine of ideas. The same applies to Popper's project, of studying ideas in themselves — a very old hat indeed. One may focus on ideas without assuming that they exist separately from ideating subjects. In general, fictions are permissible as long as they are not confused with existents. There is no hope for a metaphysical doctrine based on confusion.

Until recently, most researchers identified the gray matter in the cerebral cortex as the organ of the mind — as Hercule Poirot, Agatha Christie's hero, keeps saying. The recent discovery that white matter too participates in the mental is only a minor correction: from a philosophical point of view, what matters is that the mental is material, and that only nerve cells organized into specialized but interrelated systems, like the hippocampus, can have mental functions beyond memory. Sculptors have the choice of material, from mud and wood to marble and bronze, but nature does not. Thinking matter cannot be made with liver cells or lung cells: only neurons and glials qualify for that function. And evolution would not have happened without cellular diversity: only embryonic cells are pluripotent. In the real world type of stuff is of the essence, so all the functionalist talk about "multiple realizability" is false.

In conclusion, total or systemic materialism does not suffer from the limitations of earlier versions of materialism, in particular physicalism, biologism, computerism, and dialectical materialism. Thus, systemic materialism is the same as scientific materialism. It is also the philosophical crown of scientism — the subject of the next chapter.

CHAPTER 10

SCIENTISM

Scientism is the thesis that *all cognitive problems are best tackled by adopting the scientific approach*, also called "the scientific attitude" and "the scientific method." While most contemporary philosophers reject scientism, arguably scientists practice it even if they have never encountered the word.

10.1 Scientism Misunderstood and Slandered

The embryologist and biophilosopher Félix Le Dantec (1912: 68) popularized the word 'scientism.' And Lalande's (1939: 740) classical *Vocabulaire* defined the corresponding concept in clear terms, namely as "the idea that the scientific spirit and methods should be expanded to all the domains of intellectual and moral [social] life without exception."

However, the scientism concept had been hatched much earlier in the radical wing of the French Enlightenment. And both word and concept occurred in other contexts, particularly in religious publications, where it was used in its pejorative acceptation. Peter Schöttler (2013: 98) found that, around 1900, the words 'science' and 'scientism' were usually accompanied by the following epithets in the relevant French literature: abstract, anti-religious, bankrupt, cold, dogmatic, Durkheimian, exaggerate, false, German, gross, heavy, lame, materialist, narrow, pedantic, positivist, pretentious, rationalist, secularist, socialist, stupid, and vulgar. A contemporary study might yield a similar result: after one century, science and scientism continue to be two of the *bêtes noires* of the obscurantist party.

Scientism has often been equated with positivism, in particular Comte's. While it is true that Comte stated that sociology (a word he coined) ought to be rendered scientific, he made no contributions to it, and he did not appreciate Condorcet's much earlier essays in mathematical social science. Moreover, Comte believed that sociology and biology should test their hypotheses by comparison rather than experiment. Worse, in line with the phenomenalism of Hume and Kant, he condemned all talk of atoms and the innards of stars.

Consequently, for all his praise of science, Comte's positivism can hardly be regarded as scientific. This is why Émile Meyerson (1931) — one of the two philosophers who corresponded with Einstein — missed no occasion to criticize Comte's ban on all the research projects that, like atomism and astrophysics, looked underneath phenomena.

Friedrich Hayek (1952), who, in line with the Austrian tradition, disliked the French Enlightenment, ignored the classical definition recalled above, and offered his own idiosyncratic one: scientism would be "the attempt to ape the natural sciences" in social studies. This slanted concept of scientism is the one that has prevailed in the humanities, particularly since the post-modernist counter-revolution that started around 1950. This reactionary trend recruited those left behind, as well as those who blame science for the sins of "the establishment." To understand this change in the evaluation of scientism, we must take a closer look at its historical background.

10.2 Enlightenment Scientism

Along with secularism, egalitarianism, humanism, and materialism, scientism was a component of the radical wing of the French Enlightenment, from Diderot, Helvétius, d'Holbach and La Mettrie to Cloots, Condorcet, Maréchal, Mirabeau, and Paine. This strand was at odds with both the moderate wing of the same vast cultural movement (d'Alembert, Montesquieu, Rousseau, Turgot, and Voltaire) and the far smaller and timid Scottish Enlightenment — Hume, Smith, and Hutcheson. (See Israel 2010 for the differences between the two wings.)

Whereas the above-mentioned French were revolutionaries both philosophically and politically, albeit of the armchair kind, the Scots were reformists. In particular, the moderates did not share the atheism, egalitarianism, and republicanism of the French and American radicals. Nor did they adopt the scientistic manifesto in Condorcet's reception speech at the French Academy in 1782. There he declared his trust that the "moral [social] sciences" would eventually "follow the same methods, acquire an equally exact and precise language, attain the same degree of certainty" as the physical [natural] sciences (Condorcet 1976).

Condorcet's scientism did not involve the ontological reductionism exemplified in recent years by sociobiology, pop evolutionary psychology, neuroeconomics, and the rest of the purely programmatic neuro hype. Indeed, in the same lecture, Condorcet noted that in the moral [social] sciences "the observer himself forms part of the society that he observes." Therefore, presumably he would have welcomed the so-called Thomas theorem, according to which in social matters appearance is real, in that people react not to external stimuli but to the way they "perceive" them. So, Condorcet's scientism was not naturalistic: he knew that machines and social systems, though material rather than spiritual, are artificial, hence just as *unnatural* as science, ethics, and the law. (For the differences between naturalism and materialism, see Bunge 2009b.)

Much the same applies to Condorcet's philosophical comrades in arms, in particular d'Holbach, who treated the two branches of factual science in two different volumes: *Système de la nature* (1770) and *Système social* (1773). Their scientism was methodological, not ontological, which is why it is wrong to call it 'methodological naturalism', the way Popper (1960) did. Incidentally, the French Enlightenment was a blind spot of his, as of the entire Austrian cultural tradition: Austria had missed the Renaissance, the Reformation, the Scientific Revolution, and the Enlightenment, and only in mid-19th century it leaped from the Middle Ages to its own Industrial Revolution and the "Late Austrian Enlightenment" marked by Bolzano, Mendel, Mach, and Boltzmann.

Besides, Popper, never eager to define his key words, in particular 'historicism', 'collectivism', 'rationality' and 'scientism,' had left social philosophy to Hayek, on whom he depended to be hired by the London School of Economics, and who "managed to corrupt his socialism," as Hacohen (2000: 486) has documented. For all of these reasons, Popper should not be taken as an authority on either scientism or social science.

The Vienna Circle adopted all of the principles of the radical wing of the French Enlightenment except for realism and materialism: it remained shackled to the phenomenalism essential to Hume, Kant, Comte, Mach, and Duhem, according to which all there is (or at least all that can be known) is appearance (to someone). With the exception of Otto Neurath, the Circle was indifferent to social science, which on the whole paid at least lip service to the Enlightenment's scientistic tradition: this is what their *unified science* program meant (Neurath 1955).

The neoclassical economic theorists, in particular Jevons, Menger, Pareto, Walras, and Marshall, had practiced scientism in the pejorative sense of the word: theirs is best called *mock science*. Indeed, they produced a voluminous body of work, namely neoclassical microeconomics, bristling with symbols that intimidated the non-mathematicians but were neither mathematically well-defined nor empirically supported (Bunge 1996; 1998; 1999a). In particular, they did not subject their hypotheses to empirical tests, the way Daniel Kahneman and the Zürich group of experimental economics headed by Ernst Fehr have been doing in recent years — alas, with bad results for economic orthodoxy (see, e.g., Gintis *et al.* 2005).

10.3 Counter-Enlightenment Anti-Scientism

The German philosopher Wilhelm Dilthey (1883), who was heavily indebted to both Kant and Hegel, as well as to biblical hermeneutics, wrote the most influential anti-scientism manifesto. This early hermeneutic text had both an ontological and a methodological component. The former consisted in the thesis that everything social is *geistig* (spiritual, moral) rather than material. Its methodological partner is obvious: the social studies are *Geisteswissenschaften* (spiritual sciences), hence deserving a method of their own. This was *Verstehen*, or comprehension, or interpretation, rather than explanation in terms of mechanisms and laws.

According to Dilthey, *Verstehen* consists in the intuitive or empathic "understanding" of an actor's feelings and thoughts — what contemporary psychologists mistakenly call "theory of mind". The tacit reasoning underlying Dilthey's view is this. According to vulgar opinion, history is the doing of a few Great Men — mostly warriors and geniuses. Hence one must empathize with them, or put oneself in their shoes, if one hopes to understand what has been going on. *Verstehen* consists in empathy or fellow feeling (*mit-gefühl*) according to Dilthey, and in guessing intentions or goals in the case of Max Weber (see Bunge 1996). What Husserl (1931) meant by "the meaning of Thing" is anyone's guess.

Hence, according to those philosophers, the need to do *verstehende* (interpretive) or "humanistic" rather than scientific studies. Of course, neither Dilthey nor his followers suspected that the problem of "inferring" (guessing) mental states from behavior is an inverse problem, and as such one for which no algorithms are available, and that any proposed solution to it is speculative and dubious (see Bunge 2006).

It is usually assumed that Max Weber was the most famous of the practitioners of "interpretive sociology," the subtitle of his magnum opus (Weber 1976). Besides, he regarded himself as a follower of Dilthey's "logic" (Weber 1988). But, at least since his admirable defense of objectivism or realism (Weber 1904), Weber tried to practice the scientific method, and occasionally he even adopted historical materialism; for instance, when he explained the decline of ancient Rome not as a result of moral depravation, as we were told at school, but of the shrinking of the slave market, which in turn resulted from the cessation of the expansionary wars, the main source of slaves (Weber 1924). In short, Weber started out his sociological career as an opponent of scientism, only to become an occasional if inconsistent practitioner of it (Bunge 2007). By contrast, his rival Emile Durkheim (1988) was all his life a vocal defender and practitioner of scientism — and as such the butt of much of the anti-scientistic rhetoric of his time.

Hermeneutics, or textualism, is an offshoot of Dilthey's thesis that communication is the hub of social life. His followers, such as Claude Lévi-Strauss, Clifford Geertz, Paul Ricoeur, and Charles Taylor, held that societies are "languages or like languages." Hence the study of society should concentrate on the symbolic, and aim at catching "meanings," whatever these may be. (In colloquial German, *Deutung* may denote either sense or intention — an equivocation that facilitates the jump from the goal of an agent to the meaning of his utterances.)

But of course if one focuses on words, rather than needs, wishes, habits, and objective constraints, one cannot understand why people work, cooperate, or fight. No wonder that hermeneutics had nothing to say about the main social issues of our time, from world wars to technological unemployment to the rise of the US or China, to the empowering of women and the continued subjection of the so-called developing countries to the powerful ones (see Albert 1988 on the uselessness of hermeneutics in social science).

For example, in 1966, while the hermeneuticist Clifford Geertz (1973) was thinking about the "meanings" of Balinese cockfighting, General Suharto, supported by the U.S. government, ordered the execution of at least 500,000 of his Malaysian supporters of President Sukarno, one of the leaders of the bloc of non-aligned nations.

On the other hand, a scientistic social science, one focusing on objective facts, from rainfall to harvest, rather than on beliefs and rites, and armed with statistics instead of literary similes, should have much to say about social processes and how to steer them.

10.4 Testing Anti-Scientism

How has the interpretive or humanist approach fared? Let us evaluate the pivotal theses of the anti-scientism movement, from Dilthey's *Verstehen* to mid-20th century hermeneutics or text interpretation.

The natural/cultural dichotomy was stillborn. Indeed, by the time Dilthey proclaimed it in 1883, a number of hybrid sciences had already been in existence, notably anthropology, human geography, psychophysics, epidemiology, and demography. And shortly thereafter further biosocial sciences emerged, among them medical sociology, physiological psychology, developmental cognitive neuroscience, social cognitive neuroscience, and socioeconomics — though not biopolitics.

For example, explaining such bottom-up processes as *Puberty → Altered feelings → Changed social behavior*, and top-down ones like *Subordination → Higher corticoid level → Lower immunity → Sickness*, call for the merger of neuroscience, cognitive neuroscience, and sociology. Such a merger of disciplines is a clear breach of the natural/cultural divide decreed by Dilthey's school.

The preceding examples should refute the charge that scientism involves microeducation or leveling down. When accompanied by a science-oriented ontology, scientism favors the merger or convergence of different disciplines rather than microreduction (Bunge 2003b). All such disciplinary mergers show is that the nature/culture wall erected by Kant and inherited by the interpretive or humanistic school hinders the advancement of science, which is divergent in some cases and convergent in others.

The **Verstehen** *method has been fruitless.* Indeed, no interpretive (or humanistic) student of society has ever come up with true conjectures about any important economic, political, or cultural processes, such as the rise and corruption of democracy. In particular, no hermeneuticist has explained the rise of totalitarianism or the decline of empires during the last century. Worse, the ablest members of this school, the interpretivist Max Weber and the phenomenologist Max Scheler, supported the German empire during World War I.

However, a few students of society outside the scientific camp have produced some insightful work. Suffice it to recall the brilliant essays of Thorstein Veblen, Norberto Bobbio, and Albert O. Hirschman. In addition, Margaret Mead, Clifford Geertz, Napoléon Chagnon, and Colin Turnbull have written popular albeit controversial descriptions of certain exotic customs. However, none of these anthropologists was particularly interested in ordinary life except for sex, play, or war; their subjects seemed to subsist on thin air. (See Harris's 1968 and Trigger's 2003 explicitly realist and materialist counterweights.)

To see social studies at their best one must look at the work of anthropologists, archaeologists, sociologists, and historians of the scientistic persuasion, such as the *Annales* school, Gunnar Myrdal's monumental and influential *American Dilemma*, the inventory of archaeological pieces before being drowned by the Aswan dam, and the massive study *The American Soldier*. The publication of the latter work in 1949 elicited the anger of the humanistic school, and it also marked the coming of age of the scientific wing of American sociology, with Robert Merton at its head and the *American Sociological Review* as its flagship.

Why has anti-scientism failed? Arguably, it failed because it was blind to the big picture and condemned the scientific method, inherent in all of the scientific achievements since the Scientific Revolution. Moreover, when tackling new cognitive problems, every contemporary investigator takes scientism for granted, as will be argued anon.

10.5 The Philosophical Matrix of Scientific Research

Most philosophers take it for granted that science and philosophy do not intersect: that scientists start from observations, or from hypotheses, and handle them without any philosophical preconceptions. A glance at the history of science suffices to indict this thesis as a myth. A quick examination of a few open problems will corroborate this harsh verdict.

Let us imagine how a scientist would tackle an open problem, such as (a) whether "dark matter" and "dark energy" defy all known physical laws; (b) which if any acquired characters are inheritable; (c) whether some animals can be in conscious states; (d) how to manage in a scientific manner social systems such as business firms and armies; and (e) whether the courts of law can and should use scientific evidence, such as DNA sequencing, in addition to the traditional detection methods like finger-printing and witness interrogation.

Would our scientist refuse to investigate these problems, joining Noam Chomsky and his fellow "mysterians" (radical skeptics), in holding that matter and mind are mysterious and will forever remain so?; would she jump into *medias res* instead of starting by reviewing the relevant background knowledge?; would she fantasize about anomalous events and abnormal or even supernatural powers, or would he filter out the spiritualist fantasies?; would she remain satisfied with listing appearances or symptoms, or would she conjecture possible patterns and their underlying mechanisms?; would she remain satisfied with hunches, or would she seek empirical corroboration?; would she confine her attention to the object of her research, or would she place it into its context or wider system?; and would she dismiss out of hand all concerns about the possible harmful effect of her findings?

Admittedly, all of the previous questions are loaded. But this is the point of our exercise: to suggest that genuine scientists do not adopt or even investigate the first guess that comes to mind, just as they do not question at once all of the antecedent knowledge.

Let us see how a scientistic student is likely to tackle the five problems listed above.

Is "dark matter" anomalous or just little-known matter? The only way to find out what whether it exists and what it is, is to use the known theoretical and experimental tools, to catch samples of it and try to detect some of its properties. At the time of writing this is a "hot" question, and there is growing consensus that dark matter, if it exists, is the debris left by cosmic rays when going through ordinary matter rather than tiny black holes, as had been conjectured earlier. Stay tuned.

Was Lamarck right after all? In recent years, genetics and evolutionary biology have been enriched with epigenetics, the newest branch of genetics, which has shown conclusively that some experiences cause the methylation of the DNA molecule, an inheritable change. This discovery did not vindicate Lamarck: it only showed that the Darwinian schema (mutation-selection) can come in more than one version. (See, e.g., Szyf *et al.* 2008).

Can animals be in conscious states? The popular literature is full of anecdotes about consciousness in animals of various species. But anecdotes are not hard scientific data. Some of the best such data have recently been obtained by effecting reversible thalamic and cortical inactivations — procedures that are beyond the ken of the "humanistic" psychologists. It turns out that there is mounting evidence for the hypothesis that animals of various species can be conscious (e.g., Boly *et al.* 2013).

Can social systems be scientifically managed? Operations Research, the most sophisticated phase of management science, was born overnight from the multidisciplinary team put together at the beginning of World War II by the British Admiralty to face the great losses inflicted by the German submarines on the merchant navy that was transporting food and ammunition to England. The problem was to find the optimal size of a naval convoy. The mathematical model built by the said team, led by the physicist Patrick Blackett, showed that size to be middling, large enough to justify air coverage but not so large as to invite a fleet of enemy submarines — a result that must have baffled the economists who love to maximize. The navy accepted this contribution by a handful of newcomers to military strategy, and the naval losses decreased. This result encouraged business experts to construct mathematical models for similar problems, such as finding the optimal size of stocks ("inventories"). Thus scientism scored another victory over the traditional or humanistic party, this time in the field of sociotechnology.

Can the law become scientific? In recent years, criminology and jurisprudence, as well as their practice in the courts of law, have benefited from biology, psychology, and sociology (see, e.g., Wikström & Sampson 2006). Indeed, DNA testing is now admissible in the courts, juvenile criminal justice is slowly changing as we learn that the adolescent frontal cortex is not yet fully mature, and criminal law as a whole is changing as the social causes of crime are being unveiled and the rehabilitation techniques are being perfected. All of these advances are accomplishments of scientism.

All five problems are currently being investigated on the scientistic assumption that the scientific method is the royal road to objective truth and efficiency in all of the scientific and technological fields. Moreover, in all five cases more than scientism is being presupposed: realism, materialism, systemism and humanism too are being taken for granted. For instance, the study of animal consciousness assumes (a) the *realist* hypothesis that mental processes in the experimental animals are real rather than figments of the experimenter's imagination; (b) the *materialist* thesis that mental states are brain states; (c) the *systemic* principle that the problem under study, like all of the Big Questions, is part of a bundle of problems to be tackled anatomically as well as behaviorally; and (d) the *humanist* injunction to respect animal welfare — which in turn suggests refraining from prodding at random the animal's brain just to see what happens.

If scientific research indeed presupposes the philosophical theses that characterize scientism, then this view does not oppose the humanities, as is often claimed. What the proponents of scientism oppose is the antiscientific stand adopted by Hegel, Schopenhauer, Nietzsche, Dilthey, Bergson, Husserl, Heidegger, the Frankfurt school, and the postmodernists who have taken possession of the humanities in France. Do those enemies of rationality deserve being called 'humanists' if we accept Aristotle's definition of 'man' as "the rational animal?"

10.6 What's So Special About Science?

Laymen, and even some famous philosophers, have felt offended when told that science is not just "refined common sense:" that only the scientific method has given us counter-intuitive pieces of knowledge, such as that sunlight is a product of nuclear reactions in the sun's bowels; that we descend from fish; that our remote ancestors were not mighty hunters but feeble gatherers and scavengers; and that brain-imaging can detect traces of some experiences.

Philosophers have used such examples to argue against their ordinary-language and phenomenalist colleagues, and in favor of the idea that science

Background knowledge → *Cognitive Problem* → *Research*
    ↑   ↓
Updated fund of knowledge ← *New knowledge item*

Fig. 10.1.   The positive feedback mechanism of the growth of science.

begins where common sense stops, because most things and events are imperceptible, so we must conjecture them.

Scientific research works best at imagining objective or impersonal truths because it matches both the world and our cognitive apparatus. Indeed, the world is not a patchwork of disjoint appearances, as Hume, Kant, Mach, the logical positivists and the many-worlds metaphysicians believed, but a system of material systems. Besides, humans can learn to use not just their senses, which yield only shallow and often misleading appearances, but also their imagination, as well as to check it through observation, experiment, and compatibility with other items in the fund of antecedent knowledge (Bunge 1967b).

Besides, unlike its alternatives, science can and does grow through positive feedback, a mechanism whereby some of the output is fed back into the system. See Figure 10.1.

However, science does not come cheap: the continuance of its growth requires spending close of 3% of the national GDP on research and development (Press 2013).

In short, the support of science and adherence to scientism have repaid handsomely, economically as well as culturally, whereas betting on obscurantist philosophies threatens the growth of knowledge — a process that has been going on, albeit with temporary setbacks, since the Scientific Revolution.

CHAPTER 11

TECHNOLOGY, SCIENCE, AND POLITICS

Everyone uses the word 'technology,' but not everyone means the same by it. Some people identify it with engineering, others with an assemblage of tools and machinery, and still others with the specialized knowledge used for making or altering things in a rational way (see Agassi 1985, Quintanilla 2005). We shall understand 'technology' in the latter way. That is, we define 'technology' as *the body of knowledge used to make or change things with the help of science.* In other words, whereas scientists study reality, technologists design, repair, or maintain artifacts.

11.1 Defining and Placing Technology

Although technology requires an ever-growing science input, it is not just applied science. Indeed, a creative technologist is endowed with a technological intuition and know-how that are uncommon among scientists. This is why very few scientists have held patents, whereas many inventors have not taken science courses.

Just as good gardeners are said to possess a green thumb, we may say that original technologists are blessed with a gray intuition, which allows them to imagine the outline of a device that goes from desired output to the requisite input and input-output mechanism. Just think of all the widely used artifacts invented without using any scientific knowledge, from the plough, the pump, the carpenter's tools and the ballpoint to the windmill, the bicycle, the typewriter and the first airplane.

Most basic scientists do not invent anything useful because they are just not interested in utility. Shorter: scientists tackle cognitive problems, and occasionally they obtain solutions that help solve technological problems, but science alone is insufficient to generate technology. For example, the engineer and entrepreneur Guglielmo Marconi used the electromagnetic waves, first theorized by the great James Clerk Maxwell, to launch the radio and build an industrial empire. Only Maxwell had "seen" the said waves in his couple of mathematical triplets, and only Heinrich Hertz devised and built emitters and receptors of such waves, whereas Marconi exploited those results of basic research.

It is conceivable that other engineers too would eventually come up with the first radio receiver. In fact, the Croatian Nikola Tesla and the Russian Aleksandr Popov preceded Marconi by a few years, but they lacked his wealth, business acumen, and showmanship. Much the same holds for Thomas A. Edison, Bill Gates, and Steve Jobs.

In line with our previous definition, modern technology is coeval with science, which involves typically modern ways of thinking and doing, in particular devising mathematical models and laboratory setups. The crafts, from making flint arrowheads to cooking and writing, were invented and practiced without the help of science, hence they should not be included in technology. By contrast, modern agronomy and veterinary, computer lore and advertising, dentistry, and criminal law belong in technology, for they use results of basic research. For example, management science uses results of psychology.

Increasingly since mid-19th century, new technology has resulted from deliberate "translations" of results of basic research. For example, pharmacology is applied biochemistry, in particular the result of searches for new molecules with a *possible* medical use. Applied scientists look for new truths, just like their basic-science colleagues, but they are likely to do their work at medical faculties or pharmaceutical companies rather than at faculties of science, and their research is often paid for by commercial or military concerns, because of its merely possible commercial value.

Science and technology nurture one another and they are characteristic of a dynamic culture, just as dogmatism is the signature of a stagnant or dying culture. Note that I am using the sociological concept of culture as the system composed of producers and users of symbolic, or cognitive, moral, and artistic, items. This concept differs from the one introduced by German idealism into anthropology. According to the latter, everything

Demand → Background knowledge → Practical problem → Research
↑   ↓
Social change ← Manufacture ← Prototype ← New artifact design

Fig. 11.1.   The positive feedback mechanism of technological growth.

social is cultural by virtue of being spiritual — whence the name *Geisteswissenschaft*, or cultural science, for social science. Second caution: the East/West tension is not a "clash of cultures," as Samuel Huntington claimed, but an aspect of the conflict between imperialist powers and resource-rich countries. Oil wars cannot be disguised as disputes over the niqab, or even over palm dates.

The growth mechanism of technology is a positive feedback, just like that of basic science, but it is moved by social (or antisocial) needs and wishes rather than by sheer curiosity. In other words, technology starts and ends at society, not at the fund of knowledge. See Figure 11.1.

A society without technology is definitely premodern, and one without original technology is backward even if it imports artifacts produced elsewhere. Thus, technology is one of the engines of contemporary culture, along with science, the arts, and the humanities. By contrast, religion, which was the center of medieval culture in the West, and is still central in Islamic countries, belongs in premodernity. Remember what the Ayatollah Khomeini told the Italian journalist Oriana Fallaci in 1979: the Islamic Republic will permit the importation of Western artifacts, but not that of science, for it is inimical to religion.

11.2 Technology and Science as the Engines of Modernity

The earliest journal about our subject, founded in 1959, is *Technology and Culture*. This name suggests that technology interacts with culture instead of being a part of it. Kant and Hegel took no notice at all of technology, and engineers were not invited to literary *salons*. Even Karl Marx, a technophile and historian of technology, was unsure whether to place it, and even science, in the material infrastructure or in the spiritual structure. He admired technology both for freeing workers from the hardships and indignities of manual work and for its contribution to the large-scale production of affordable goods, but not for its rich intellectual and artistic content. Marx seems to have fallen for the economist's fallacy, that science is only the handmaiden of industry. And Engels felt contempt for Newton, whom he called an *Induktionsesel* (inductive ass).

Among classical philosophers, only Descartes and Spinoza respected craftsmen. And, the radical wing of the French Enlightenment exalted the crafts and engineering to the point of devoting to them a good portion of the *Encyclopédie* edited by Diderot, d'Holbach, and initially also d'Alembert. Not even the Scots Adam Smith and David Hume, who admired steam-powered machines for being labor-saving devices, placed engineering in culture, perhaps because they regarded it as only sophisticated craftsman-ship. (For the conceptual wealth of modern technology, see Bunge 1985b; Raynaud 2016.)

The above-mentioned philosophers, as well as the postmodern scrib-blers about what they call *technoscience*, would have been astonished if informed that modern technology makes intensive use of advanced science, including abstract mathematics. Yet this is common knowledge among engineering students. In particular, electronic engineers, nanotechnologists, and roboticians have to learn a lot of theoretical classical mechanics, classical electrodynamics, electron theory, solid state physics, and the underlying quantum mechanics, as well as some of the new knowledge produced in physics labs.

11.3 Technoscience?

Technology is so dependent on science that sometimes they are fused together, and the product of this fusion is called *technoscience*. But the dif-ferences between the partners in question are just as obvious as their com-monalities. Their differences become obvious when comparing a scientific research project with a "translational" one.

For one thing, whereas scientific research aims at truth, technological development aims at utility. This is why scientific theories, unlike tech-nologies, are not patentable, and why private firms do not sponsor research in astronomy, paleontology, anthropology, or historiography. For another, while scientific theories are tested for truth, technological designs are tested for utility. Besides, in principle science is international, whereas the advanced technologies are useless in preindustrial countries, which require "appropriate" technologies. Finally, whereas science is morally neutral, technology is morally partial, since some technologies are beneficial whereas others are harmful, and still others are just as ambivalent as the proverbial knife.

Fortunately, Hitler and his gang did not distinguish science from technology, and charged their greatest theoretical physicist with the task of manufacturing the German bomb. But Heisenberg had no idea about it, and apparently was not even interested in it — so much so, that he took a long holiday in Hungary, where he read and wrote philosophy. By contrast, the Americans understood that their Manhattan Project would be a gigantic endeavor generating a new technology and requiring a new management style.

They appointed General Leslie Grove, an able administrator and ruthless politician, along with Oppenheimer as the scientific director. Their team quickly swelled to more than 500,000 employees and delivered on the goods — the two bombs that destroyed Hiroshima and Nagasaki, and told the world which the new top dog was. In short, the leaders of the Manhattan Project did not swallow the technoscience story concocted by philosophers ignorant of both partners of this bicephalous creature.

11.4 Technophilia and Technophobia

There are two mains attitudes toward technology: blind acceptance, or technophilia, and rejection, or technophobia. And either can be adopted either moderately or fanatically. Most people in the advanced societies admire technological advances regardless of their negative effects on everyday life (such as increasing sedendarism and its medical concomitants) and on the environment (such as pollution).

By contrast, most technophobes give no reasons for their opposition to innovation, whether technological or of some other kind, just because they remain shackled to the past, warts and all. Thus the so-called feudal socialists of the early 19th century objected to capitalism because it introduced unemployment and cut the lord-serf links that had ensured the social immobility of the traditional societies; theirs was just a case of political conservatism, which in turn favors old privilege. The contemporary technophobes are also afraid of the possible social changes brought about by radical and pervasive technological innovation. These technophobes can be either religious, like the theologian Jacques Ellul, or secular, like the existentialist Martin Heidegger, who admired the Nazi war machine while continuing to criticize technology in general just for being modern.

Nowadays the most visible technophiles are those who claim that there are technologies capable of counterveiling the negative side of technological innovation. The best-known case of this kind is that of the economists who claim that geoengineering, which exists only in their imagination, could regulate climate. Another case of the same kind is that of the moral philosophers who discovered that global warming could be avoided by lowering the height of people by exactly 15% — surely a trivial task for geneticists.

Another bizarre case of dogmatic technophilia is that of the famous mathematician John von Neumann, who announced that we are on the eve of the "essential singularity," or the time after which automata will design all the inventions. Von Neumann's prestige was such, that in 2008 he inspired the foundation of the Singularity University, supported by NASA and by some big corporations. Obviously, none of the participants in this technofiction adventure remembered that robots idle unless activated by programs designed by flesh-and-blood people. Nor did they wonder about the moral issues raised by their dystopia.

11.5 The Moral and Political Aspects of Technology

Basic scientists could not harm anyone even if they wanted. Harming, just like doing good actions, takes practical skills and raises moral issues. Such issues emerge every time "translations" of basic knowledge are envisaged. Just think of the translations of parts of biochemistry into either a tool to improve crops or to kill with poison gas.

A tragic example of such moral ambivalence is that of the eminent chemist Fritz Haber, who had become instantly famous for inventing the process for manufacturing ammonia from atmospheric nitrogen and hydrogen ($N_2 + H_2 \rightarrow 2NH_3$). Seized with patriotic zeal, Haber also invented the toxic gases used in World War I, as well as Zyklon B, later used in the Nazi death camps. In 1915, when news of the success of his gas in the second Yprès battle came, Haber threw a party for his military and civilian handlers. The morning after, his beautiful and accomplished wife and colleague Clara Immerwahr shot herself with her husband's service handgun. Unrepentant, Haber immediately departed for the Eastern front to supervise gas warfare. He got the Nobel Prize in 1918, but two decades later was forced to emigrate for being Jewish.

The ambivalence of both ammonia and Haber is atypical. The "translations" from lab to factory or battleground are hard to accomplish because few scientific products are deadly, and few brains are driven either by curiosity or by utility with equal intensity. These difficulties explain why the private laboratories have produced far less science and technology than the universities (Raynaud 2015).

A few big companies, like Bell, IBM, Dupont, IBM, and IG Farbenindustrie, have employed some scientists, but mostly as occasional consultants, much in the way a big publisher may ask a great novelist to evaluate a submission, though never to write a masterpiece. The prospect of immediate reward only produces potboilers. Great original works, whether artistic, scientific, or technological, come only from creative passion.

The largest technological enterprise, the Manhattan Project (1939–47), which manufactured the first nuclear bombs, employed nearly all the American physicists and many British ones as well, but it produced no memorable scientific findings. Its only product served only to frighten everyone around the world and to launch the first world power in history.

In conclusion, (a) unlike basic research, which is autonomous or self-propelling, technology is heteronomous or other-pulled: it has aims other than the advancement of knowledge for its own sake; (b) whereas basic science is morally neutral, technology can be good or bad; and (c) whereas scientific research produces cultural goods, technology, though a cultural enterprise, produces merchandises.

These conclusions confute the pragmatist and Marxist opinions on the relations between action and knowledge, in particular between industry and science. The same conclusions support the science and technology policy that strengthens basic research and resists the privatization of public universities as well as their submission to ideologies hostile to science.

Even the most conservative American leaders have understood the benefits of investing 2.8% of their GDP in scientific research. By contrast, the European Union invests only 1.8% of its GDP in the same activity, although Western Europe has yet to recover the level it had attained before

World War II. It has been suggested that this disparity is partly due to the influence of postmodern nonsense, in particular the so-called Continental philosophy, born and bred in France and Germany.

11.6 Genuine and Bogus Knowledge

Rita Levi-Montalcini, who won a Nobel Prize for discovering the nerve growth factor, titled her wonderful memoirs *In Praise of Imperfection*. Her point was that scientists start a research project when they realize that they ignore something that may turn out to be interesting — that is, when their curiosity is piqued. By contrast, the know-it-all types, such as the fanatics and the believers in total ideologies, are happy reading old books and commenting on them.

For example, whereas Aristotle and Marx knew that only original research produces new genuine knowledge, their dogmatic followers seek to know about their old mentors instead of tackling new problems with the help of state-of-the-art methods. Unsurprisingly, such limited curiosity produces useless knowledge in the best of cases, and bogus knowledge in the worst. Let us peek at the two more influential instances of pseudo-knowledge: religion and pseudoscience.

Although religion it is still popular and politically influential, it does not belong in *modern* culture, for it does not stimulate the renewal of genuine knowledge. If it did, we would have religious chemistry and engineering, along with neoliberal logic and epistemology. But nobody writes eloquent epistles to the Corinthians any more. In short, from being firmly entrenched in culture, religion has become utterly marginal to it. There are still a few religious scientists, but they do not include their dogmas in their research projects. Moreover, the Catholic Church has leaned to admit evolutionary biology and socioeconomic heresy.

The only field where groundless superstition and uncontested authority still rule is pseudoscience. Recent cases of wild speculations advertised as bold scientific discoveries are its-from-bits physics, many-world cosmology, and the theory of everything — all of which are still I.O.U.s. Worse, the first of them contradicts all the well-corroborated theories that assume that things, unlike bits, have energy, and the unborn theory of everything overlooks the large differences among things belonging to levels of organization,

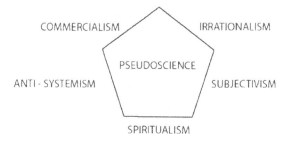

Fig. 11.2. The conceptual matrix of pseudoscience.

like the physical and the social, which are very distant from one another. As for the its-from-bits fantasy, it overlooks the fact that information, far from being a universal entity or property, is a feature of sophisticated information systems that include items like codes and decoders, as well as emitters and receivers, which are artifactual and were inaccessible to our remote ancestors.

Anyone can learn about religions or pseudosciences, but neither of them accounts truthfully for anything, whether natural or artificial: one can claim to be familiar with theology, but not with God. Both fields belong in the realm of fantasy, where no rigorous research is ever conducted. And, like any other conceptual system, every piece of bogus knowledge is born inside a large matrix, but one that differs significantly from that of science: see Figure 11.2. The scientists who refuse to place their own work in the philosophical matrix sketched in Figure 3.1 risk falling for some pseudoscience. In other words, science is not enough to protect us from myth: only a scientific philosophy can help. But this antidote is hard to detect, and even harder to get. Come and join the lab!

11.7 Science and Philosophy: An Odd Couple

The science–philosophy couple has undergone many radical changes since their birth as twins a few millennia ago. They were one for the ancients, but they split when Justinian proclaimed the superiority of Christian theology and banned the pagan works. Under any theocracy, whether Christian, Islamic, Jewish, or other, philosophy, if tolerated at all, was to be *ancilla theologiae*. Modernity secularized and distinguished philosophy from science. Philosophers speculated and argued, whereas scientists observed, measured, experimented, and formalized their hypotheses.

For example, the first modern scientists, such as Galileo, did not mix their astronomical observations with their philosophical thoughts, let alone their religious heterodoxies. The same applies to Galileo's near-contemporaries Harvey and Vesalius. For them, science and philosophy were not only distinct but also mutually indifferent, although, of course, they took rationality, realism, materialism, and systemism for granted.

The final breach between philosophy and science occurred only toward the late seventeenth century, when John Locke (1690) had the gall of writing a treatise on human knowledge without having studied the scientific revolution that had just happened during his lifetime in his own country, namely the birth of Newtonian mechanics and astronomy. (The two men met only in their old age and in their capacity of high-ranking civil servants.)

Since then, philosophy and science have coexisted in mutual ignorance. Ignorance did not prevent Hume from criticizing Newtonian mechanics, Berkeley from ignoring it, Kant from attempting to improve on it adding a repulsive force to balance gravitational attraction, or Goethe from blaming Newton for having used a prism to decompose pure white light into rays of all colors. Nor did ignorance prevent Hegel from claiming that Kepler's laws entail Newton's law of motion, Berzelius for proposing the first explanation of chemical reactions — while at the same time proclaiming that philosophy dominated science. At about the same time, Comte inverted the science–philosophy connection, asserting the primacy of science, but he wished to confine it to phenomena, and consequently he condemned the atomic theory and astrophysics.

Marx sang praises to natural science, but rebuked the Darwinians for rejecting Hegel's dialectics. He also argued, this time correctly, against the view that evolutionary biology could explain social change; he criticized *avant la lettre* the sociobiology that would become popular one century later. Schopenhauer wrote that the will keeps the world moving, while Nietzsche coined the slogan *Fiat vita, pereat veritas!*, and Dilthey denied the posssibiliy of a social science.

Just when biology was taking the intellectual culture by storm, and starting to have a big impact on medicine, Henri Bergson — rewarded with a Nobel Prize — claimed that science could not explain living things, and wrote against the special theory of relativity. Croce, Gentile, Husserl, and their followers, as well as Wittgenstein and his school, just ignored all the scientific breakthroughs of their time. The only philosopher–scientists of the nineteenth century were John Stuart Mill, Karl Marx, Ernst Mach, Charles Sanders Peirce, and Ludwig Boltzmann. However, this is not the place to evaluate their philosophical contributions.

Coda

The disconnection between philosophy and science has been disastrous for both. In fact, it favored the birth of Hegel's and Schelling's philosophies of nature, and it has tolerated contemporary pseudosciences such as rational-choice theory and psychoanalysis; as well, it has slowed down the progress of psychology and social science. The said disconnect has also permitted the multiplication of pseudoproblems (such as the ravens and the grue paradoxes) as well as of academic industries such as many-worlds metaphysics. The latter, which specializes in zombies, brains in vats, life without water, and superhuman computers, has diverted the attention of philosophers from the genuine problems generated by the study and control of the one and only real world, such as synthetic life, genetic determinism, free will, environmental protection, and the possibility of expanding democracy to include all the sectors of society (see Bunge 2001; 2009a).

To sum up, most modern philosophers have either asserted the mutual independence of science and philosophy, or have held that one of them should dominate the other. My own stand on the matter may be condensed into the slogan *Philosophize scientifically, and approach science philosophically* (Bunge 1957).

APPENDIX 1

FREEING FREE WILL: A NEUROSCIENTIFIC PERSPECTIVE

Agustín Ibáñez*,†,‡,§,**, Eugenia Hesse*,
Facundo Manes*,§ and Adolfo M. García*,¶

*Instituto de Neurociencia Cognitiva y Traslacional (INCyT)
Fundación INECO, Universidad de Favaloro
Consejo Nacional de Investigaciones Científicas y Técnicas (CONICET)
Buenos Aires, Argentina
†Universidad Autónoma del Caribe, Barranquilla, Colombia
‡Center for Social and Cognitive Neuroscience, School of Psychology,
Adolfo Ibáñez University Santiago, Chile
§Australian Research Council Centre of Excellence
in Cognition and Its Disorders, Australia
¶Faculty of Elementary and Special Education (FEEyE)
National University of Cuyo (UNCuyo) Mendoza, Argentina

Free will (FW) was originally conceived as a dualistic and Neoplatonic notion, and these foundational properties pervade current views rooted in cognitive neuroscience. In an attempt to foster progress beyond those traditional tenets, here we propose an unorthodox neurocognitive approach to the construct. First, we explicitly assess three traditional

---

**Corresponding author: aibanez@ineco.org.ar

assumptions that should be avoided for FW to be fruitfully explored, namely, that FW is (a) categorical in ontological terms (an all-or-nothing capacity); (b) intrinsically dependent on consciousness; and (c) rooted in deterministic or non-deterministic principles. We analyze prototypical neuroscientific claims suggesting that FW is illusory and show that these considerations rely on the three classical assumptions listed above. The boundaries and dualistic foundations of classical accounts of FW can be considered misleading, or at least non-scientifically motivated. Conversely, a renewed neurocognitive conception of FW can rest upon the following principles: (a) like several other cognitive and affective domains, FW is not an all-or-nothing faculty; (b) conscious activity underlying FW is a non-contradictory, emergent property of unconscious mechanisms; and (c) processes rooted in both determination and self-determination coexist in the neurocognitive underpinnings of FW. These reconsiderations pave the way for a new research agenda, in which FW constitutes the capacity to make flexible decisions (only some of which involve moral responsibility) and reason about ensuing consequences on the self and the environment. To conclude, we update the incipient knowledge regarding brain networks relevant to FW, and call for future research to frame it as a natural, neurocognitive, and situated phenomenon.

Keywords: Free will, cognitive neuroscience, neurocognitivization.

A1.1 The Notion of Free Will: Origins and Traditional Assumptions

"[W]hen man by his own free-will sinned, then sin being victorious over him, the freedom of his will was lost." In this passage from The *Enchiridion*, Saint Augustine set forth a conception of free will (FW) that would be prototypically assumed for centuries to come. Resurfacing a tradition with roots in Roman North Africa, under the mixed influences of Christianity and Zoroastrianism, his notion involved an opposition between good and evil and was inextricably bound to decision-making skills and moral responsibility. Saint Augustine's conception mutated, as can be seen in comparing *De Libero Arbitrio* with *Confessions*. However, his dualistic and neoplatonic premises took hold in the Western world, constraining classical accounts of FW. Today, breakthroughs in cognitive neuroscience breathe new life into the debates on the nature and conceptualization of FW.

FW is usually understood as the ability to choose among alternative courses of action. Indeed, the classical idea of "freedom of choice" hints to an individual's autonomy to consciously favor one option over several others, independently of external factors. Different versions of the law and economics posit that a person's responsibility, guilt, and merits depend on how he/she exercises this integral FW. Though somewhat less explicitly, similar conceptions abound in scientific approaches, which typically frame FW in terms of three main properties:

a. Categorical capacity: FW is usually conceptualized as an all-or-nothing attribute. An individual either acts fully out of his will or entirely lacks volition, regardless of temporal, spatial, or otherwise contextual constraints. Thus conceived, FW constitutes a *mosaic* conceptual entity: a person's actions cannot be characterized as being only partially voluntary.

b. Consciousness: Both the decisions and moral responsibilities attached to FW are assumed to rely on conscious processes. In fact, the absence of consciousness during decision making may invalidate the attribution of FW.

c. Intrinsic dualism between physical determinism and immanent will: In several philosophical trends, FW has been contrasted, asserted, or denied on the basis of deterministic principles. The minimal assertion of this position is that all events in the world are the result of previous events. The ensuing dualistic picture separating a pre-determined reality from an intentional will has led FW to be conceived as (a) an illusion, (b) a transcendental faculty which overrides physical determinism, or (c) a subjective immanence which may or not be compatible with versions of soft determinism.

These three properties are present in many lay and formal conceptions of FW. For instance, they underlie notions of guilt in the law, decisional liberty in economics, sinfulness in religion, and volition in psychology. Moreover, as will be shown below, they have also pervaded the interpretation of oft-cited results in cognitive neuroscience.

A1.2 The Resilience of the Original FW Conception in Cognitive Neuroscience

Prominent neuroscientists from the eighties to date have reedited the original debates surrounding FW (Navon 2014; Smith 2011). This recast has been fueled by experiments showing that preconscious brain activity precedes conscious decisions, alongside similar demonstrations in other cognitive domains. In brief, such evidence has been copiously interpreted as an argument against the existence of FW (Smith 2011), with authors like Daniel Wegner characterizing the construct as an epiphenomenon. To address this issue, below we focus on the foundational and prototypical experiment reported by Benjamin Libet *et al.* in 1983.

In Libet *et al.*'s (1983) study, participants indicated the moment at which they became consciously aware of the decision to move a finger. To this end, they noted the position of a moving dot on a clock when they became aware of such a decision. Although this sign of awareness preceded the actual movement by ~200 ms, a neurophysiological correlate (the readiness potential) was identifiable around 550 ms before the movement. The conclusion was that *unconscious* brain activity relevant to the task began 350 ms before subjects became *conscious* of their decision to move. The experiment has been ardently criticized on methodological and interpretive grounds,[1] and we will not repeat that criticism here. Our aim is to go beyond the caveats already identified and show that even if the experiment had circumvented all technical and theoretical shortcomings, it would still be limited by its epistemological foundations.[2] In particular, the

---

[1] Critics of Libet's seminal experiment (e.g. Gomes 2002; van de Grind 2002; Verbaarschot *et al.* 2015) have pointed out numerous infelicities, including experimental flaws (e.g., non-replicable effects, time precision issues, faulty measures of awareness) and conceptual limitations (e.g., inadequate consideration of time as a causal effect, different conclusions for the same results, generalization problems). Moreover, the experiment does not offer a single measure of the actual decision time; rather, it considers the behavioral outcome of a metacognitive estimation of when the decision emerged. This is no trivial observation, given that metacognitive estimations of time can be notoriously inaccurate, especially when performed on events longer than 1–2 s (Klein 2002; Danquah 2008). In brief, what Libet's experiment shows is that metacognitive estimations of a previous decision to move tend to occur after an objective modulation of the brain's motor potential timeline.

[2] Indeed, similar conclusions have been set forth by authors who managed to overcome some of the methodological caveats in Libet's experiment.

conception of FW it incarnates assumes the three classical properties we have listed above.

First, the study frames FW as categorical faculty. The above results can be taken as a refutation of the very existence of FW only if the latter is conceived as a spatiotemporally integral entity. For this view to be entertained, one must further assume that the estimation of the exact movement-onset time is free from bias. Otherwise, the study cannot even be said to address FW at all, as it would only measure retrospective metacognitive estimations of time perception. The estimation of when one decides to move can be used as a proxy of (an illusory) FW only if it is considered categorically correct and absolute. This stands in contradiction with the fact that people can make good estimations in some situations and very bad ones in others; for example, we are better at estimating time and other quantities when sufficiently rested than during periods of fatigue (and the same holds true for decision making and event recollection). Moreover, this view also assumes that FW is absolute in a spatial sense. If FW is not present when we estimate the timeline of our conscious decision to move, then FW is missing altogether from our cognitive repertoire. It has been proposed that the allegedly "illusory" nature of FW present in Libet's experiments may actually be generalized to all kinds of subjective experiences (Libet 2006). But how could one anticipate and guarantee the success of this generalization?

Second, Libet's interpretation assumes consciousness as a prerequisite of FW: since the "unconscious" readiness potential precedes awareness of (the estimation of) one's decision, FW lacks the intrinsic property of consciousness and, consequently, does not exist (Smith 2011). Such a conclusion rests on two assumptions: (a) FW should be initially and emphatically conscious, and (b) unconscious and conscious processes are categorically opposed, there being no chance of them becoming intermixed. For Libet, the presence of unconscious cerebral processes prior to a subjective experience is enough to challenge the existence of FW (Libet 2006). As we will describe below, this two-fold argument runs counter to current neuroscientific views of unconscious and conscious processes (Nahmias 2015). In addition, the assumption that the readiness potential causes the conscious decision seems to incarnate the *post hoc ergo propter hoc* fallacy. Indeed, causal attributions do not follow from mere temporal successiveness. We cannot even assume that earlier and later events necessarily follow a chain of neural causalities coming from the same "source events." Temporally aligned patterns may result from functionally independent brain locations. Thus, interpretations of Libet's results manifest two main flaws: the inexistence of FW cannot be adduced from the unconscious nature of a neural event, and the latter cannot be taken as the necessary cause of an ensuing conscious action.[3]

Third, classical interpretations of Libet's experiment are rooted in a dualistic separation between physical determinism and an immanent will. If FW is determined by a biophysical event (the readiness potential, which in principle can be reduced to biological rules, and linked to physical realm), then FW is an illusion because all decisions would be preceded by previous physical events. If this were true, we could predict willing actions based on existing biophysical priors in the real world. Going beyond the Libet experiment, if we want to assume the FW illusion, full-fledged determinism calls for a complete and absolute theory of biophysical realms. We would need to possess the eye of God, granting us complete knowledge of all relevant objects, actions, forces, and events in order to reduce any future event to a set of pre-existing factors (Smith 2011). Of course, at present no single theory or master equation can integrate all previous conditions and predict a person's decision in a given situation. Even when we can reduce these factors to a few critical ones and correctly predict subjective states, it does not mean that FW itself can be predicted from biophysical, psychological, or cultural processes. Absolute determinism (rooted in reductionism) may be available to God, but not to science.

Radical reductionism is an atypical position in science and it proves intrinsically incomplete, even within well-established fields such as physics or chemistry. One could thus uphold FW as a real entity even in the absence of a complete theory of it. In addition, current trends in evolution, thermodynamics, quantum mechanics, logic, and mathematics have rejected absolute determinism. All these disciplines have shown that even when an emergent domain can be predicted from its previous conditions, prediction does not necessarily mean determinism, and even deterministic

---

[3] Notice, also, that Libet's conception is blind to the vast granularity of the notions of consciousness and FW. The operations underlying finger movement are hardly the same as those framing less trivial decisions, such as getting married or moving abroad. Decisional and/or volitional processes in each case are insurmountably different.

rules can be intrinsically unpredictable. In other words, even in an onto-logically deterministic world, our theories (and their related facts and predictions) will be always probabilistic and rooted in non-radical reductionism. No current theory of FW (or of any other cognitive faculties, for that matter) can be totally reduced to a set of previous constituents. Otherwise, we should deal with a dualistic conception of FW à la Libet, in which the "cerebral mental field" can be correlated with cerebral events but is non-physical by definition (Libet 2006).

In sum, mainstream perspectives on FW in contemporary neuroscience have done little to circumvent the three properties long adopted by classical philosophical accounts. However, this is not because the field lacks theoretical and empirical tools to forge alternative viewpoints. Below we sketch one such reconceptualization, challenging the three properties under discussion.

A1.3 A New (Neurocognitive) Outlook on FW

From a neuroscientific perspective, FW can be conceptualized as the capacity to make flexible decisions (only some of which involve moral responsibility) and reason about ensuing consequences on the self and the environment. FW would thus constitute a complex, high-level adaptive faculty supported by various sub-processes, including conscious and unconscious decisional operations as well as individual differences in relevant domains (reasoning skills, moral cognition, emotional regulation, social emotions). As other complex affective-cognitive processes, FW could thus only be understood as non-mosaic neurofunctional system engaged in constant bidirectional exchanges with other domains.

If FW is a product of brain function, then it must be subject to the general principles governing neurocognition, including non-linear relationships with other systems, non-discrete levels of activation states, and convergences among conscious and unconscious processes during situated activities. Once we accept that FW constitutes a neurocognitive phenomenon, there is little point in describing it via tenets other than those assumed for well-characterized systems, such as memory, emotion, or language. Hence, we shall argue that a plausible neuroscientific conception of FW *does not require* a commitment to (a) the all-or-nothingness principle, (b) the categorical differentiation between conscious and unconscious processes, or (c) a strict antagonism between determinism and non-determinism.

A1.3.1 *FW is not an all-or-nothing faculty*

Picture Jorge Luis Borges at a library, sitting next to a Chinese student of Spanish who has been taking lessons for six months. One may ask: which of them knows Spanish and which one does not? But the answer would be trivial, as it follows from an ill-posed question. Indeed, Borges did not know everything there is to know about Spanish (every word of every dialect, every grammatical pattern of every period in the historical development of the language), and neither is the Chinese student completely ignorant of Spanish (he can order a coffee, follow basic instructions, construct new sentences). Rather, we could ask more informative questions: *How* does each of them know Spanish? Or *how much* do they know? This implies moving from binary ontological speculations toward an operationalization of the construct in the quest of descriptive and explanatory insights. The same maneuver can be applied to reconceptualize FW.

As is the case with other complex cognitive domains, such as classical or social decision making, moral cognition, emotional processing or memory, FW need not be framed in terms of an all-or-nothing faculty. The existence of memory is not at odds with the fact that people forget things. Neither is the existence of decision making or moral cognition challenged by the recognition that individual or situational predispositions may at times bias both domains. If FW is conceived as a neurocognitive system, why should it be discussed in terms of an ontological dichotomy? FW may be limited, extended, or reduced by specific situational variables, related cognitive processes, or even physiopathology. However, rather than casting doubts on the existence of FW, these modulations bring evidence on the critical neurocognitive mechanisms associated with it. This non-mosaic conceptualization also assumes that FW can be constrained by temporality (it may expand or recede at different moments), sociocultural factors, and individual differences. Yet, just like our inability to efficiently recall events in stressful situations does not entail the inexistence of memory, neither do transient or ever-present automatic processes preclude the existence of FW.

A1.3.2 *As a non-contradictory, emergent property of unconsciousness*

Consciousness plays a central role in several conceptions of FW (Nahmias 2015; Shepherd 2012). Nevertheless, contemporary neuroscience assumes that consciousness cannot be understood without unconsciousness. Both forms of neurocognitive processing interact profusely in our daily ponderings, feelings, and actions. There is no principled reason to postulate that FW should be the exception. Indeed, our willing and conscious decision to perform a manual movement is subtly modulated by unconscious motor-semantic coupling effects triggered by preceding verbal information (García & Ibáñez 2016). The willing bodily event is neither fully conscious nor fully unconscious: it is a mixture of both. More generally, reducing the debate on FW to an opposition between consciousness and unconsciousness leads to a categorical error, a metonymic explanatory leap, or a simplistic heuristic strategy — explaining a phenomenon by reference to only one of its variables.

Note, also, that the spatiotemporal granularity at which one explores a given domain determines emerging perspectives on it. During spontaneous dialogue, in a scale of seconds or minutes, myriad communicative instances can be detected which respond to the interactants' volition. However, if one considers the intervening processes in a scale of milliseconds, critical dynamic changes could be detected which operate below the threshold of willing control. And it seems counterproductive to ask whether deliberate processes triggered those below our conscious control or vice versa. A scientific inquiry into FW deserves more sophisticated treatments than "chicken or egg" dilemmas.

Consciousness can actually be understood as an emergent property of unconscious operations, probably based on neurocognitive integration mechanisms which allow us to explicitly focus on a given inner process. For example, integration-to-bound models of decision making identify the initial intention to act as the dynamical interaction of unconscious and conscious workings (Murakami *et al.* 2014). These ideas actually lie at the heart of mainstream theories in cognitive neuroscience, including the global workspace model, neurodynamic models of consciousness, or body awareness models, to cite some representative examples (Baumeister *et al.* 2013; Craig 2009; Lau & Rosenthal 2011; Seth *et al.* 2006). None of these leading trends in neuroscience is compatible with binary, dualistic conceptions of mental faculties. Indeed, even those scholars who consider that research into FW is a non-scientific enterprise agree that the construct must be framed in flexible rather than binary terms (Montague 2008). By the same token, we propose that FW may be more aptly conceived as a continuum including myriad partial and gray areas which influence one another. Thus, willing actions and decisions necessarily require several unconscious processes which tacitly inform conscious deliberations (Roskies 2012).

A1.3.3 Determination and self-determination as co-existing constraints of FW

From a pure *explanans* perspective on FW, a radical resolution of the antagonistic dilemma between determinism and indeterminism is valid only if we have a scientific model capable of anticipating all the events of the world (or the mental world, in this case). Full identity between *explanans* and *explanadum* is attainable only through the eye of God. This possibility escapes not only construals of FW, but the whole of science as we know it.

From a neurocognitive perspective, FW is both determined and self-determined (Nahmias 2012), following the same principles that apply to memory, moral cognition, or decision making. As an emergent process, FW involves self-determined capacities to orchestrate different processes leading to a deliberate decision. The music of the FW orchestra depends on instruments such as decision mechanisms (probability, uncertainly, risk), cognitive flexibility (working memory, inhibition), moral emotions (guilt, shame, pride), and reasoning, among many others. The orchestra's self-organization underlies the self-determination of FW, but this does not mean that the music is limitless. The neurocognitive instruments and musicians can become worn out by disease, situational conditions, or cultural constraints. Importantly, the orchestra can only produce the music that available instruments and musicians are able to generate. The possibilities of this orchestra are neither unlimited nor perfect, but this does not deny the existence of the system's inner music.

A1.4 Toward a Non-Mosaic View of FW Supported by Network Science

How should a novel cognitive neuroscience of FW deal with the above considerations? First, complex cognitive processes are not dependent on a single region *per se*. In fact, there is no single brain area causing the FW. As a unitary object, FW only can be understood at a conceptual or analytical level. From a neurocognitive perspective, its underlying processes are submergent, emergent, and reticularly organized phenomena. Second, various brain regions indexing different FW-related processes (decision making, moral cognition, reasoning, conscious states) impact on the neural correlates of FW proper (Roskies 2010). Contemporary neuroscience has moved from isolated, homuncular, and single-mechanism explanations toward an emergent, network-based picture of the neurocognitive activity — especially for complex cognitive domains. Of course, we currently lack complete theories of any and all human faculties. If this does not preclude progress in studies on memory, language, or social cognition, why should more stringent demands be placed on FW? Even without a complete theory of FW we can study how different brain regions and processes impact on this faculty. For example, mild electrical stimulation of an isolated area (the anterior cingulate cortex) elicits specific FW-like responses within associated distributed networks, both cortically and subcortically (Parvizi *et al.* 2013).

We are still far from a fully convergent model of this faculty. FW impacts on extremely complex social contexts, and is influenced by several factors, such as adaptive strategies, preferences, reward evaluation, reinforcement learning, social cooperation, competition, and control, as well as other parameters such as uncertainly, ambiguity, or probability. Methodologically speaking, multiple studies in animals and humans have assessed FW-related processes through behavioral, pharmacological, and lesion-based approaches, as well as via EEG, fMRI, model-based fMRI, PET, and TMS recordings. In the absence of integrative frameworks to jointly interpret such a vast empirical domain, our approximations to the neurocognitive basis of FW will remain at best partial in the foreseeable future.

Nevertheless, some consensus can be traced in recent works (Nahmias 2012; Roskies 2010; 2012; Zhu 2004). An extended neural network mainly related to monoaminergic frontostriatal loops and limbic loops, in interaction with other domain-specific networks, may be critical for FW (note that this network includes the lateral, medial, and orbitofrontal cortices, the striatum, the amygdala, the insula, the basal ganglia, the anterior cingulate cortex, and monoaminergic pathways). Also, different processes related to FW should activate different subcircuits of this broad network; e.g., individual decision-making networks. In addition, social decision making seems to involve areas associated to social cognition and integration of choices, whereas simple volitional tasks can engage pre-supplementary and parietal areas during non-social decision making. Other domains, such as executive functions, memory, emotion, and feelings can modulate the neural correlates of FW. Moreover, a number of selected domains more directly related with some aspects of FW, such as volition (Roskies 2010) or moral decision making (Christensen & Gomila 2012), imply extended and overlapping neural systems.

In brief, FW can be related to a very wide network, including more specific neural substrates for its critical and associated sub-functions. These considerations regarding network diversity, overlap, and non-specificity may sound very preliminary and problematic. Nevertheless, they have not been here posited for FW in an *ad hoc* manner; on the contrary, they are central to current and well established neurocognitive models of others complex domains, such as a contextual social cognition (Ibañez & Manes 2012), moral cognition (Moll *et al.* 2005), volition (Roskies 2010), or wisdom (Meeks & Jeste 2009).

A1.5 Conclusions

Some renowned neuroscientists have argued that we need critical neuroscientific experiments to demonstrate whether FW actually exists. Here, we propose a different conceptual challenge: the boundaries and dualistic assumptions behind classical accounts of FW can be considered illusory, or at least non-scientifically motivated. A naturalization (or, more precisely, a neurocognitivization) of FW, freed from previous assumptions of absoluteness, consciousness, and dualistic tensions, will offer new insights into the notion, building on tenets which robustly characterize other neurocognitive domains, such as decision making and moral cognition.

A new, post-classical, post-ontological, and post-dualistic program could allow us to test specific hypotheses on the nature and dynamics of FW as a natural, neurocognitive, and culturally situated phenomenon. Crucially, however, this endeavor should consider the current limits of neuroscience to deal with complex processes, especially when these involve private, subjective, non-transferable inner experiences (Roskies 2010). Experimental philosophy has much ground to cover in the pursuit of a scientific program to study FW (Nahmias 2012; Nichols 2011). The theoretical platform sketched in this chapter could inspire new experimental approaches to free FW research from the constraints of traditional (and scientifically fruitless) ontological speculations.

So far, science has offered more reliable knowledge about external and distant objects, such as stars and planets, than it has about the inner universe within our brains. Much like the very notion of FW, this idea was also introduced early on by Saint Augustine. In his *Confessions*, he wrote: "And men go abroad to admire the heights of mountains, the mighty waves of the sea, the broad tides of rivers, the compass of the ocean, and the circuits of the stars, yet pass over the mystery of themselves without a thought." Yet, just as we have renewed our conception of FW, so can we gain revealing insights into our inner workings by asking the right questions (and avoiding ill-posed ones). Chances are that Augustine himself would not be convinced about the prospects of a neurocognitive approach to FW. We hope that, in time, breakthroughs rooted in this new conception of FW will allow us to pay more concrete tribute to this giant on whose shoulders we stand.

Acknowledgments

This work was partially supported by grants from CONICET and the INECO Foundation.

References

Baez, Sandra, Blas Couto, Teresa Torralva, Luciano A. Sposato, David Huepe, Patricia Montañes, *et al.* 2014 Comparing moral judgments of patients with frontotemporal dementia and frontal stroke. *JAMA Neurology* 71(9): 1172–1176.

Baez, Sandra, Philip Kanske, Diana Matallana, Patricia Montañes, Pablo A. Reyes, Andrea Slachevsky, *et al.* 2016. Integration of intention and outcome for moral judgment in frontotemporal dementia: Brain structural signatures. *Neurodegenerative Diseases* 16(3–4): 206–217.

Baumeister, Roy F., E.J. Masicampo & Kathleen D. Vohs. 2011. Do conscious thoughts cause behavior? *Annual Review of Psychology* 62: 331–361.

Christensen, Julia F. & Antoni Gomila. 2012. Moral dilemmas in cognitive neuroscience of moral decision-making: A principled review. *Neuroscience and Biobehavioral Reviews* 36(4): 1249–1264.

Craig, A.D. (Bud). 2009. How do you feel — now? The anterior insula and human awareness. *Nature Reviews Neuroscience* 10(1): 59–70.

Danquah, Adam N., Martin J. Farrell & Donald J. O'Boyle. 2008. Biases in the subjective timing of perceptual events: Libet *et al.* 1983 revisited. *Consciousness and Cognition* 17(3): 616–612.

García, Adolfo M. & Agustín Ibáñez. 2016. A touch with words: Dynamic synergies between manual actions and language. *Neuroscience & Biobehavioral Reviews* 68: 59–95.

Gomes, Gilberto. 2002. The interpretation of Libet's results on the timing of conscious events: A commentary. *Consciousness and Cognition* 11: 221–230.

Ibañez, Agustin & Facundo Manes. 2012. Contextual social cognition and the behavioral variant of frontotemporal dementia. *Neurology* 78(17): 1354–1362.

Klein, Stanley A. 2002. Libet's temporal anomalies: A reassessment of the data. *Consciousness and Cognition* 11(2): 198–214; discussion 314–325.

Lau, Hakwan & David Rosenthal. 2011. Empirical support for higher-order theories of conscious awareness. *Trends in Cognitive Sciences* 15(8): 365–373.

Libet, Benjamin. 2006. Reflections on the interaction of the mind and brain. *Progress in Neurobiology* 78(3–5): 322–326.

Libet, Benjamin, Curtis A. Gleason, Elwood W. Wright & Dennis K. Pearl. 1983. Time of unconscious intention to act in relation to onset of cerebral activity (Readiness-Potential). *Brain* 106: 623–642.

Meeks, Thomas W. & Dilip V. Jeste. 2009. Neurobiology of wisdom: A literature overview. *Archives of General Psychiatry* 66(4): 355–365.

Moll, Jorge, Roland Zahn, Ricardo de Oliveira-Souza, Frank Krueger & Jordan Grafman. 2005. Opinion: The neural basis of human moral cognition. *Nature Reviews Neuroscience* 6(10): 799–809.

Montague, P. Read. 2008. Free will. *Current Biology* 18(14): R584–R585.

Murakami, Masayoshi, M. Inês Vicente, Gil M. Costa & Zachary F. Mainen. 2014. Neural antecedents of self-initiated actions in secondary motor cortex. *Nature Neuroscience* 17(11): 1574–1582.

Nahmias, Eddy. 2012. Free will and responsibility. *Wiley Interdisciplinary Reviews Cognitive Science* 3(4): 439–449.

——. 2015. Why we have free will. *Scientific American* 312(1): 76–79.

Navon, David. 2014. How plausible is it that conscious control is illusory? *The American Journal of Psychology* 127(2): 147–155.

Nichols, Shaun. 2011. Experimental philosophy and the problem of free will. *Science* 331(6023): 1401–1403.

Parvizi, Josef, Vinitha Rangarajan, William R. Shirer, Nikita Desai & Michael D. Greicius. 2013. The will to persevere induced by electrical stimulation of the human cingulate gyrus. *Neuron* 80(6): 1359-67.

Roskies, Adina L. 2010. How does neuroscience affect our conception of volition? *Annual Review of Neuroscience* 33: 109–130.

——. 2012. How does the neuroscience of decision making bear on our understanding of moral responsibility and free will? *Current Opinion in Neurobiology* 22(6): 1022–1026.

Seth, Anil K., Eugene M. Izhikevich, George N. Reeke & Gerald M. Edelman. 2006. Theories and measures of consciousness: An extended framework. *Proceedings of the National Academy of Sciences of the United States of America* 103(28): 10799–10804.

Shepherd, Joshua. 2012. Free will and consciousness: Experimental studies. *Consciousness and Cognition* 21(2): 915–927.

Smith, Kerri. 2011. Neuroscience vs. philosophy: Taking aim at free will. *Nature* 477(7362): 23–25.

Soon, Chun Sion, Marcel Brass, Hans-Jochen Heinze & John-Dylan Haynes. 2008. Unconscious determinants of free decisions in the human brain. *Nature Neuroscience* 11: 543–545.

van der Grind, Wim. 2002. Physical, neural, and mental timing. *Consciousness and Cognition* 11(2): 241–264.

Verbaarschot, Ceci, Jason Farquhar & Pim Haselager. 2015. Lost in time...: The search for intentions and Readiness Potentials. *Consciousness and Cognition* 33: 300–315.

Zhu, Jing. 2004. Locating volition. *Consciousness and Cognition* 13(2): 302–322.

THE PHILOSOPHY OF MIND NEEDS A BETTER METAPHYSICS[1]

Martin Mahner

Zentrum für Wissenschaft und kritisches Denken
GWUP e.V., Rossdorf, Germany

A2.1 Introduction

In examining the relationship between brain and mind, the philosophy of mind refers to mental properties, mental states, mental events, etc. It also uses concepts such as "identity," "causation," "supervenience," or "emergence." Thus the philosophy of mind is full of metaphysics, but it has no fully fledged metaphysical theory, let alone a generally accepted one. In addition, the metaphysical notions used in the philosophy of mind are often based on ordinary language concepts rather than scientific ones. This is unfortunate, because in my view this slows down the progress of the field in that the same old problems, if not pseudoproblems, keep being discussed over and over without much hope of resolution. For example, both the (in) famous zombie argument against materialism, and the functionalist claim that computers or other machines could develop consciousness, simply dissolve in certain ontologies.

It should be interesting, therefore, to introduce a promising metaphysical theory, and to see why these examples are non-problems in the light of such

---

[1] Reprinted from Miller, Steven M., ed. (2015) *The Constitution of Phenomenal Consciousness: Toward a Science and Theory*, pp. 293–309. Amsterdam: John Benjamins, with permission by the author and the publisher.

a metaphysics. As analytic metaphysics has been a thriving philosophical field for quite some time, there are lots of approaches from which to choose.[2] The in my opinion most promising ontology does not exactly belong to the philosophical mainstream, which is why it is worthwhile to examine its power here: the ontology developed by Mario Bunge (1977; 1979), who applied his approach to the mind–body problem early on (Bunge 1979; 1980; Bunge & Ardila 1987), without, however, exploring all its possible ramifications and consequences. As I have applied Bunge's ontology to the philosophy of biology before (Mahner & Bunge 1997), and as I have summarized his metaphysics in a German book (Bunge & Mahner 2004), I shall borrow from these works whenever convenient.

A2.2 The Materialist Metaphysics of Mario Bunge

A2.2.1 *Things and properties*

Figure A2.1 illustrates the logical structure of Bunge's ontology — a structure which I shall follow in this section.

Bunge's ontology is a so-called substance metaphysics, which considers the notion of a material thing as the most fundamental concept. (By contrast, process metaphysics regard processes as more fundamental than things.) In the style of Aristotelian metaphysics, the notion of a (material) thing is developed from the concepts of property and substance. A substance or bare individual is that which "bears" or "carries" properties. Note that neither properties nor substances are real in the sense of having autonomous existence: there are neither self-existing properties nor self-existing substances; there are only propertied substances, that is, things. Thus properties and substances precede things only analytically, not factually.

We may distinguish several types of properties. The first distinction is between intrinsic and relational properties. An *intrinsic* property is one that a thing possesses independently of any other thing, even if acquired under the influence of other things. Examples: composition, electric charge, mass,

---

[2] Note that in this chapter I use 'metaphysics' and 'ontology,' as well as the corresponding adjectives, synonymously. This is in tune with the philosophical tradition that introduced the term 'ontology' as a name for a de-theologized metaphysics. In my view it is regrettable that nowadays 'ontology' is often just used in the sense of the union of the reference classes of scientific theories, that is, as that which tells us "what there is."

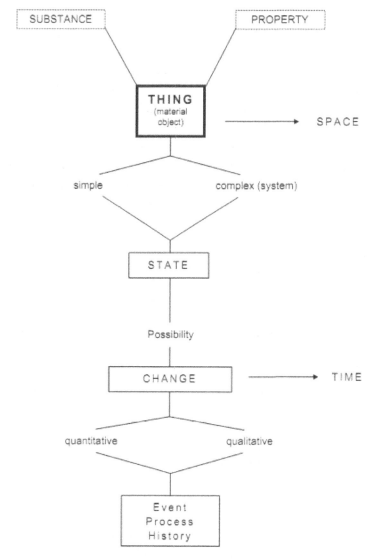

Fig. A2.1. The basic structure of Mario Bunge's ontology (modified from Bunge & Mahner 2004). The figure is to be read from top to bottom, and the nodes are to be understood in the sense of a logical or definitional precedence, that is, a given concept at a certain level is defined with the help of the preceding concept(s). For example, the concept of property is logically prior to the concept of a state, which in turn is prior to that of an event. The arrows towards the concepts of space and time indicate that they are not basic, but derived from the concepts of thing and change. That is, being a material or concrete thing is not defined in terms of spatiotemporality.

consciousness. *Relational* properties, by contrast, are of course properties that a thing has only in relation to some other thing(s). Examples: speed, weight, descent. A special type of relational property is the so-called secondary or phenomenal property. The prime example is of course color. Whereas reflected wave length (or, more precisely, the distribution of spectral reflectances) of something is a primary property, color is a secondary property, that is, wave length (or, more precisely, spectral reflectance distribution) as represented by some organism equipped with suitable sensory organs and a sufficiently complex nervous system. In this construal secondary properties are relational in that they are possessed by the subject/object system rather than by either the subject or the object alone. The object has only primary properties, and if a subject has a phenomenal experience without actually perceiving (representing) an outer object, it is either dreaming or hallucinating.

Another distinction (also going back to Aristotle) is the one between essential and accidental properties. Essential (constitutive) properties are those that determine the nature or essence of a thing. They are those that we need to describe of what kind a thing is. For example, the difference between my brain and a rock is given by the essential properties of both. An accidental property, by contrast, makes no difference to what a thing is. For instance, whether a quartz crystal is located in Australia or in Africa makes no difference to its being a quartz crystal. These examples indicate that essential properties come in clusters: they are *lawfully* related. This implies an ontological construal of laws, not an epistemological one in the sense of law statements. Laws or, more precisely, lawfully related properties are what law statements refer to, if true. Ontic laws in this sense are *inherent* in the things which possess them. They are neither contingent nor imposed from without. Laws belong to the nature of things. Not even God could change the lawful behavior of things without changing their nature.

Thus, Bunge is a representative of what in the philosophy of nature is called the "new essentialism" (Ellis 2002), which may better be called *scientific* or *nomological essentialism*.[3] This essentialist view of laws as properties

---

[3] As is well known, essentialism is not exactly going strong these days, in particular in the philosophy of biology, which has become rather anti-essentialist. For example, due to the enormous variation of organisms, many philosophers of biology believe that there are no laws (= law statements) in biology. But this does not entail that organisms do not behave lawfully: it is just that it often makes not much

of things is very important, because a thing's lawfully related properties determine what is *actually* possible for a thing, as opposed to what is just *logically* possible.

Furthermore, there are qualitative and quantitative properties, as well as manifest and dispositional ones, where dispositions can be either causal (e.g., the disposition of a glass to break) or stochastic (e.g., the propensity of an atom to decay).

Finally, we have a type of property that is most relevant to the philosophy of mind: systemic (or emergent or supervenient) properties as opposed to merely resultant ones. Most things are complex, that is, they are composed of parts, which may in turn be composed of further parts: they are systems. A property that only a system as a whole possesses, yet none of its parts in isolation, is a *systemic* property; otherwise, a property is *resultant.* If I assemble a computer in a factory, its parts have a mass, and so has the final product. The mass of the whole is just the (additive) result of the (quantitative) mass property of its parts. By contrast, the various properties that only the correctly assembled computer as a whole displays — most conspicuously its specific functions — are its systemic properties. We may as well call these systemic properties 'supervenient properties' or 'emergent properties.' Bunge prefers the term 'emergent,' which he defines in purely ontological terms: as a property of a whole that is not present in its parts. 'Emergent' is often defined in epistemological terms, namely as a property of a whole that cannot be explained or predicted from the knowledge of its parts. Yet whether or not a systemic property can be explained (or predicted) by reference to the parts of a system is immaterial for its being a (new) property of a whole. Nonetheless, the systemic properties of a whole do lawfully depend on the (essential) properties of its parts (the so-called base properties). This is why, contrary to the belief of the functionalists, your green cheese is never going to think or have feelings: its parts lack the relevant base properties.

---

sense to try to find general, let alone universal, law statements because their reference class is rather small, holding only for some subspecies, variety or even smaller units, for example; that is, only for those organisms sharing the same lawful properties. In other words, biological kinds often have only a small number of specimens (more on laws in biology in Mahner & Bunge 1997; Ellis, 2002). Consequently, anti-essentialism in biology and its philosophy is just as misleading as it is everywhere else, such as in the philosophy of mind.

To illustrate the importance of lawfully related essential properties and emergence, let us take a look at an example (Figure A2.2). A thing x has two properties P and Q, which are lawfully related, or in other words, related by the law L_{PQ}. A more complex thing y may consist of some things of the same kind as x, but possesses in addition a new (emergent) property R. If R is an essential property, it must be lawfully related with either P or Q or both. That is, y must possess at least one new law L_{PR} or L_{QR}, or perhaps even both. As a consequence, if y fails to have one or even both of the base properties P and Q, there will be no lawful emergent property R. Replace Q by a different property S (which means you replace the part x with a different part z), and you will obviously get neither R nor L_{RQ} but at most some different property T and perhaps a new law L_{ST}.

A thing is individuated (or identified) by the set of its properties at a given time. These properties are individual properties in the sense that only the given thing possesses them. No other thing can possess *my* mass or *my* age, although many other things also have a certain mass or age. We can

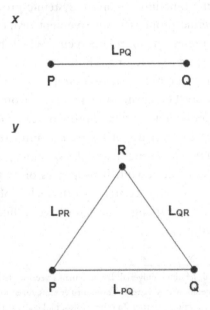

Fig. A2.2. The lawful relationship between essential properties in thing x, and the emergence of new laws in the new thing y. Explanation in the text (modified from Bunge & Mahner 2004).

thus distinguish general (or universal) properties from individual ones.[4] If things share a number of general properties, they are of the same type or kind, but each token of a given type is an individual which is individuated by its particular properties.

Finally, we should emphasize the difference between properties and predicates. Properties are always properties of concrete things, whereas predicates are conceptual representations of properties. Thus, whereas real things possess properties, only the model things in our mind have predicates. Evidently, some of our representations may be wrong in that some predicates do not represent real properties. We sometimes hypothesize that a certain thing has a certain property only to find out later that our hypothesis is false. Moreover, there are two kinds of predicates that never represent real properties. These are negative and disjunctive predicates. The absence of a property is not a property of a thing, even though it is often convenient for us to describe some things by the absence of some property. For example, as a mammal, I do not possess wings, but this does not imply that the absence of wings is a property of mine. Rather, the relevant property is that I possess forearms of a certain structure that allow me to grasp, touch, etc. Negation is *de dicto*, not *de re*. The same holds for disjunctive predicates. For example, 'heavy' or 'transparent' is not a property of anything, which can only have the property of being heavy, of being transparent, or of being both heavy and transparent.[5]

A2.2.2 *States*

In ordinary language a state is something like a uniform phase or stage of some process. For example, one says that an object is in a state of motion, or that a person is in a state of confusion. In Bunge's ontology, however, a state is something static, and this static concept is used to define the (dynamic) concepts of event and process.

---

[4] Individual properties are often called 'tropes' in analytic ontology.

[5] A consequence of Bunge's theory of properties is that Boolean algebra cannot be used to formalize ontological concepts like "property" or "supervenience." All such attempts are doomed from the start. Real properties have the formal structure of an inf-semilattice, which is a much poorer structure than Boolean algebra. For a criticism of Kim's (1978) early analysis of the concept of supervenience in terms of Boolean algebra, see Mahner and Bunge (1997, p. 32f.).

As we saw above, everything has a number of properties. The totality of properties of a thing at a certain time determines the state of the thing at the given time. Because every property can be formalized as a mathematical function, the list of n such functions is called a state function of things of the kind concerned. That is, if we have n functions F_i, the state function F of the given thing is the list or n-tuple $F = <F_1, F_2, ..., F_n>$. The value of F at time t, i.e., $F(t) = <F_1(t), F_2(t), ..., F_n(t)>$, represents the state of the thing at time t.

The set of possible states of a thing can be represented in a state space or possibility space for the thing. This is the abstract space spanned by the corresponding state function $F = <F_1, F_2, ..., F_n>$. If, for the sake of simplicity, we consider only two properties, the corresponding state space is a region of the plane determined by the axes F_1 and F_2— see Figure A.2.3. A state space for a thing with n properties is n-dimensional.

Because essential properties are lawfully related, a material thing cannot be in all its logically possible states: its really possible states are restricted by the laws that define its kind. The subset of really possible states of the logically possible state space of the given thing is called its *lawful* or *nomological* state space S_N— see again Figure A2.3.

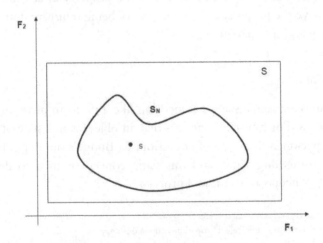

Fig. A2.3. Two properties F_1 and F_2 span a two-dimensional state space S. Any state s of the given thing is represented by a point. The laws of a material thing of a given kind restrict the logically possible state space S to a proper subset: its nomological state space S_N (modified from Bunge & Mahner 2004).

According to this construal, the set of conscious states of a brain (or of some neuronal subsystem) is a proper subset of its nomological state space. Anyway, it should be possible in principle, though perhaps not in practice, to map out the conscious state space of a brain.

Obviously, if a thing acquires a new (in particular systemic) property, we must add a new axis to its state space representation, and if it loses one, we must remove the corresponding axis. In this way, emergence can be represented by the addition of new axes to a thing's state space. The converse process, that is, the loss of properties in the course of system formation or dissolution, may be called *submergence*, and it is represented by the removal of axes.

A2.2.3 *Events and processes*

Things hardly stay in the same state. In fact they change all the time. By "change" we understand a change of state and thereby a change of the properties of a thing. Change can be illustrated by a line in a state space (Figure A2.4). Whereas a point in a state space represents a state, a line represents a sequence of states. An event can be represented as an ordered

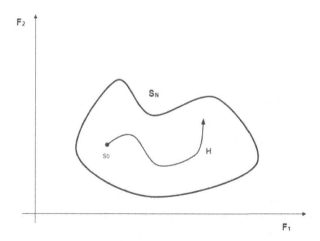

Figure A2.4. Processes may be represented by a line in the given nomological state space S_N of a thing, starting with some original state s_0. The line H is the history of the given thing between some original and some later or even final state (modified from Bunge & Mahner 2004).

pair of states <initial state, final state> or <s, s'>. As with the logically possible states of a thing, we can collect all the logically possible events in (or changes of state of) a thing to form the conceivable event space of the thing in question. And as everything can only be in nomologically possible states, it can only undergo nomologically possible changes of state, i.e., events. Thus, the set of really possible events is restricted to the nomological event space of the changing thing(s) in question.

Just as things don't stay in the same state, they usually don't just undergo singular events but sequences of such. A sequence of states or, alternatively, a sequence of two or more events of a thing is a *process* (or complex event). Thus, processes can be represented by a curve in a state space. This is important because not any old set of events is a process: only a sequence of states of one thing, however complex, qualifies as a process. Needless to say, two or more things may interact and thus form a system, whose states and changes of state can be represented in a state space of its own.

Some special types of processes are called *mechanisms*. Mechanisms are *specific* processes (characteristic functions), that is, they occur only in things of a certain kind. More in Section A2.3.

Finally, the *history* of a thing is the total set of its changes of state, from its very beginning s_0 to its end, if any.[6] See Figure A2.4.

A2.2.4 *Causes*

What Bunge calls simply 'process' is often called a 'causal process.' However, in Bunge's ontology a state of a thing is not the cause of a later state of the same thing. Youth is not the cause of old age, and the caterpillar is not the cause of the butterfly. Bunge speaks of an (external) cause only if a change of state of, i.e., an event in, a given thing generates a change of state, i.e., an event, in some other thing; or if an event in one part of a thing generates an event in another part. Thus the causal relation only holds among events.

The concept of causation can be elucidated in terms of the state space approach. Consider two different things, or parts of a thing, of some kind(s). Call them x and y, and call H(x) and H(y) their respective histories

---

[6] Bunge's ontology is therefore a specimen of what is called *endurantism*.

over a certain time interval. Further, call $H(y \mid x)$ the history of y when x acts on y. Then we say that x acts on y if, and only if, $H(y) \neq H(y \mid x)$, that is, if x induces changes in the states of y. The corresponding event e in x that brought about the event e' in y is the cause, while e' is the effect.

Just as the concept of law, the notion of causation here is ontological, not epistemological. It can be analyzed as a form of energy transfer between two things or two parts of a complex thing. It goes without saying that especially in the biological sciences many causes are hard to detect, and they require elaborate experimental and statistical methods.

If only events as defined here are causes, strictly speaking neither things nor properties are causes. In the context of the debate about mental causation (see, e.g., Robb & Heil 2008, for an overview), it is often claimed that if mental properties are real, they need to have "causal powers." But if mental properties are simply systemic properties of certain neuronal systems, they do not cause anything. (Indeed, we may consider this as a form of property epiphenomenalism.) At most the whole changing neuronal system can cause something; in other words, a neuronal system with (emergent) mental properties may behave differently than one without such properties; or, to be more precise, an organism with neuronal systems possessing emergent mental properties should behave differently than one without such properties.

There can be no mental causation without mental events, and there can be no mental events without mental things. And because, according to Bunge's emergentist materialism, there are no mental things (that is, immaterial minds in themselves), there are no mental events and hence no mental causation. While the talk of mental states and events may be a convenient shorthand in the daily business of the neurosciences, the use of metaphysically ill-conceived terms may be seriously misleading in the philosophy of mind.

A2.3 Systems and Mechanisms

Most things are not simple but complex: they are composed of other things. If a complex thing is just an aggregate of other things, such as a pile of sand, it is a heap rather than a system. Systems, by contrast, are cohesive: they have a specific structure formed by strong bonds. Except for microphysical entities, such as quarks and electrons, which are not known to be composed of further parts, virtually all things studied by science, from atoms through societies, are systems (Bunge 1979).

In the analysis of any system three aspects are important: its composition, its environment, and its structure. Bunge calls this the *CES-analysis* of systems. The *composition* of a system is of course the set of its (natural and material) parts.[7]

The environment of a system is the complement of its composition, that is, the set of all things other than the given system. However, for a scientific model of a concrete system s we do not need to take the rest of the universe into account. We can restrict our analysis to those things that may act upon s, or which s may act upon. What is relevant, then, is this *immediate* (or *causally proximate*) *environment* of s.

Finally, the structure of a system s is the set of relations among its components. Of particular relevance here are those relations that make for the cohesiveness of s. These are the *bonding* or *causal* relations. In tune with the state space approach introduced above, we can say that a relation between a thing x and a thing y is a *bonding* relation if the states of y alter when the relation to x holds. Consequently, the nomological state space of a system is not the union (or mereological sum) of the nomological state spaces of its components, but it must be construed anew taking into account in particular the emergent (lawful) properties of the given system.

The *internal structure* (or *endostructure*) of a system s, then, is the set of bonding and nonbonding relations among the components of s. Often we are interested only in the bonding endostructure of s, for example, when we want to know how neurons are connected into complex systems. As systems, or some of their parts, also interact with some external things, they do have an *exostructure* as well. Just as its endostructure, a system's exostructure is the union of the bonding and nonbonding relations among the system (or of some of its parts) and the items in its environment. Again, we are usually interested only in the bonding exostructure of a system.

---

[7] In ordinary language as well as in mereology the set of parts of a thing may include arbitrary parts which do not exist really but are the result of artificially (or conceptually) slicing up a thing into as many parts as we wish. An example are relational parts, such as the upper, middle, and lower part of a tower, whereas the natural parts of a tower are the stones, steel beams, tiles, windows, or whatever the tower is made of, as well as the further natural parts these parts consist of. The qualification 'natural' excludes such arbitrary parts, focusing on parts that belong to natural kinds. Finally, the qualification "material" excludes so-called temporal parts, which occur in perdurantist ontologies.

The notion of a bonding exostructure makes it obvious that many systems are structured hierarchically: they consist of subsystems and they are part of some supersystem(s). A scientific study of systems will therefore have to focus on some specific level of organization, that is, it will consider a system's subsystems and some of the supersystems it is part of. For example, a study of neuronal systems and their functioning needs to look not only at their molecular and cellular components but also at the whole brain, and, in the case of consciousness, at the social input of the given person. In other words, what is of interest for a scientific explanation is not the entire composition, environment, and structure of a system but only the composition, environment, and structure of some of the adjacent levels. The CES-analysis of systems is thus usually restricted to a $C_L E_L S_L$-analysis, where L refers to the relevant sub- or supersystem levels of the given system. In any case, what is quite irrelevant in the explanation of the mental in its emergentist conception is any reference to the microphysical level, because it disregards all the emergent properties of the higher-level systems in between.

As it is quite common to say that the special sciences deal with different levels of systems, such as the physical, chemical, biological, and social levels, the question arises whether there is also a mental level. In Bunge's metaphysics there is no mental level because the mental is conceived of as an emergent property of certain neuronal systems. There would be a mental level only if there were mental things above and beyond neuronal systems. So if the brain produced a self-existing mind or mental thing, such as a gland secreting some hormone, there would be a mental level. If the mental is just an emergent property existing only and whenever certain neuronal systems undergo certain processes, then there is no mental thing and hence no mental level, unless we wish to stipulate that such a mental level is just the set of all neuronal systems capable of exhibiting mental properties.[8]

The CES-analysis of systems is static. To get closer to real-life situations we need to take into account the changes of systems. For example, as consciousness is most likely a certain activity or activity pattern of highly

---

[8] Thus, the question of the "constitution of consciousness" (Miller 2007) must be understood in the sense of the (compositional, structural, environmental, and mechanismic) constitution of neuronal systems *with* mental properties in contrast to those *without* mental properties. Needless to say, it remains a most formidable task for neuroscience to distinguish these neuronal systems.

complex neuronal systems, it cannot be fully understood by a merely static analysis of neuronal systems. As we saw in Section A2.3, everything has its own nomological event space, that is, the set of nomically possible changes it is able to undergo. What is nomically possible is determined by the lawful properties of the given system, including its emergent properties. A subset of all the possible processes of a system s is the set of kind-specific processes of s. For example, many cells share a similar basic physiology, but only some cells are able to gain energy through photosynthesis including the corresponding physiological processes. These specific processes or functions may be called the *mechanisms* of the given system. A mechanism is thus a process function, not a structured thing.[9]

Note that "function" is often understood not in the sense of "process function" or "functioning" (or *modus operandi*) but in the sense of "role function". A role function, however, is something a system does with respect to some higher-level system it is part of (Mahner & Bunge 1997; 2001). For example, the role function of the heart is to pump blood. But this role function can be achieved by at least two different mechanisms or process functions: the muscular contractions of a normal biotic heart (including of course all the relevant lower-level processes) or the electric mechanics of an artificial pump replacing the heart of patients with heart failure. It seems therefore that role functions are multiply realizable, whereas process functions are not. If consciousness is a process function of certain neuronal systems, then systems of a different kind won't be able to be conscious (more on this in section 5).

In sum, taking into account the characteristic changes of systems of a given kind — its mechanisms — we may add a fourth coordinate M to the CES triple, obtaining a CESM-quadruple. Analyzing systems in terms of CESM is essentially what scientific and mechanismic explanations do.[10]

---

[9] In the literature on mechanismic explanation both a structured thing and its processes are called mechanisms (see, e.g., Glennan 2002; Machamer *et al.* 2000). It is of course correct that we talk about the mechanism of a watch, for example, but this belongs to the description of the composition and structure of a watch. It is only the specific processes that this composition and structure allows for that we call mechanisms here. Note that Bunge prefers the new adjective 'mechanismic' over 'mechanistic,' so as to avoid misunderstandings in terms of physical mechanics or machine metaphors. A social mechanism, for instance, is as far from mechanics as it gets.

[10] Although mechanismic explanation (as opposed to the classical, merely subsumptive deductive-nomological explanation) has become increasingly popular over the past two decades, in the philosophy

A2.4 Why Many Metaphysical Approaches Are Unsatisfactory

Of course there are many alternative ontological approaches. But I submit that they are more or less unsatisfactory. To see why, let's take a look at some of the metaphysical considerations of Jaegwon Kim, because he is one of the major players in the philosophy of mind, and he has also dealt with the concepts of event, substance, state etc.

Kim (1993) characterizes an event thus: "We think of an event as a concrete object (or n-tuple of objects) exemplifying a property (or n-adic relation) at a time. In this sense of 'event' events include states, conditions, and the like, and not only events narrowly conceived as involving changes" (p. 8; similarly, p. 33ff.). We further learn that "[b]y 'substance' I mean things like tables, chairs, atoms, living creatures, bits of stuff like water or bronze, and the like ..." (p. 33), and "[a] change in a substance occurs when that substance acquires a property it did not previously have, or loses a property it previously had." (p. 33). All this is the so-called "property exemplification account of events" (p. 34).

Why is this approach unsatisfactory? (a) It strikes me as odd and confusing to not regard the aspect of change as essential for the meaning of 'event.' A thing possessing a property at some time is a fact, but it is not an event. Not all facts are events. (b) An n-tuple of concrete objects is not itself a concrete object, but a mathematical representation and hence a conceptual object. Of course there are complex concrete objects, composed of many parts, but these are systems or complexes of things, forming a higher-level entity. The composition of these systems may be formally *represented* by n-tuples, but n-tuples are not out there. (c) Talk of the exemplification of properties is certainly common, but to a materialist it has a ring of Platonism to it. It sounds as if properties hovered in an immaterial realm of ideas, and once in a blue moon concrete objects instantiated or exemplified these properties. I therefore avoid any such talk. (d) Traditionally, a substance is not a concrete object, but the "bearer of properties." But as there are no such bare individuals without properties, a substance is at best an

---

of science it was proposed already back in 1967 by Bunge (1967, p. 25ff.). Unluckily, he had called it "interpretive explanation", and, unfortunately, his work was largely ignored by the philosophical community. So was his elaboration on mechanismic explanation from 1983 (Bunge 1983). Thus, it was left to others to reinvent and popularize the concept. Nowadays Peter Railton is (wrongly) credited for having invented the idea of a mechanism in 1978 (Glennan 2002, p. 343).

ontological concept, not a real thing. (e) Acquiring or losing a property is a qualitative change. Yet more frequently concrete objects change only quantitatively. For example, growing or getting older does not entail the acquisition or loss of a property, but only a change of the respective property value. Indeed, in science quantitative properties are represented as real-valued functions, so that the change of a property can be graphically illustrated as a curve in some coordinate system. A misconception of quantitative changes like these may be the reason why Kim adopts the property exemplification approach: if we consider only one general property, such as age, it seems that a concrete ageing object exemplifies the property of ageing. But this is not so: it has the general property of age, but the individual values of this property change. So we do have a change of state in this case, not just the possession of a property.

The defects of Kim's ontology provide an example for how wide-ranging the import of an ontological theory is and how important it is to base the philosophy of mind on the best available ontology, that is, an ontology that is more in tune with scientific practice and which has greater analytic power.

A2.5 Zombies and Thinking Machines

It is rather trivial to point out that philosophical views and arguments (unless they are purely formal perhaps) have explicit or tacit metaphysical presuppositions. If we accept some argument as convincing, we must then also accept its metaphysical presuppositions. Conversely, if we have no reason to accept these presuppositions, we have no reason to accept the corresponding argument. I shall argue here that in the light of the metaphysics sketched in the preceding, we have no reason to accept two well-known ideas occurring in the philosophy of mind: functionalism and the zombie problem. Functionalism is mistaken, and the zombie problem is a non-problem.

Functionalism and the associated notion of multiple realizability presuppose that matter doesn't matter because structure is all there is to mental life, so that it could occur in many different things than brains, even in artificial ones such as computers. One of the hypothetical arguments of this approach is what is called 'neuron replacement.' Imagine that we replace one neuron of a human brain with a functionally equivalent artificial electronic neuron. (The adjective 'electronic' is important here, because we are not concerned with artificially synthesized biological neurons as they would be materially equivalent with the original natural neurons.) Do the patient's brain functions and hence his mental life change after that? Probably not. Now let's continue with this neuron replacement until the entire brain consists of artificial neurons. According to the functionalists, this artificial brain would work just as well as the original one and hence display consciousness, for all that matters is the functional organization of input/output behavior, not the matter of which brains are made.

According to emergentist materialism *cum* nomological essentialism, consciousness or, more precisely, *being conscious* is an emergent property of certain complex neuronal systems undergoing certain coordinated activities. Which systemic properties are nomically possible is determined by the essential base properties of the given system's parts. This is why you cannot obtain a given systemic property out of any base properties, that is, out of parts of different kinds. What holds for the components of the system, also holds for the processes it can undergo. Being conscious is not a property of a static system, but an emergent property of a complex changing system: it involves process functions, not just some input/output role functions. But process functions are changes of state of the given system with a given composition and structure, so they cannot occur in systems that have completely different nomological state and event spaces. (It may be possible, however, to replace some role functions, such as those providing sensory input to some neuronal system capable of phenomenal consciousness, with artificial devices.)

This is why, according to our view here, multiple realizability of the mental is restricted to a variety of neuronal systems of the same kind. For example, your and my brains are most likely wired somewhat differently, so that your thinking of "2 + 2 = 4" may involve somewhat different neuronal systems and processes than my thinking of the same proposition. But this is just a variation within the same kind of matter, namely neuronal systems. Such variation in "realizability" is not multiple in the sense that it applies to *different kinds* (or even all kinds) of systems. This view is sometimes derided as neuro-chauvinism but it is a natural companion of nomological essential-ism. (So is carbo-chauvinism in the case of life.) For these (and many other) reasons, functionalism, such as in strong Artificial Life and Artificial Intelligence, turns out to be untenable in the light of this metaphysics (Kary & Mahner 2002; Mahner & Bunge 1997).

Let us turn to the alleged zombie problem. This problem rests on the view that logical possibility is all there is to possibility. If something is con-ceivable it is logically possible, and whatever is logically possible is also really possible. This approach often comes together with possible-worlds semantics, possible worlds metaphysics, modal logics, etc.[11]

Thus, we get arguments telling us, for example, that as it is not logically necessary that water is H_2O, it could be composed of different molecules in other (logically) possible worlds.[12] And here we go: because it is con-ceivable or, in other words, as it is not logically necessary that (normal) human brains have consciousness, it is logically possible that there are zombies in the sense of humans that function just as we do, but who fail to have any "mental life" at all (Chalmers 1996). So there is no necessary or lawful connection between neuronal systems and phenomenal con-scious or other mental "states," and the existence of mental properties becomes a mystery. Supposedly, the explanatory and metaphysical gap is so wide that materialism is incapable of closing it.

Now logical possibility is the only possibility that applies in logics and mathematics, but in science the relevant possibility is *nomic* or *real* possibil-ity. (Sometimes real possibility is also called 'metaphysical possibility,' but just as often it is unclear what exactly a metaphysical possibility is supposed to be.) Of course the logically impossible is not really possible, but not everything that is logically possible is actually possible. As we have seen in Sections A2.2 and A2.3, the nomologically possible state and event spaces of material things are proper subsets of their logically possible state and event spaces, respectively. And what is really possible is determined by the lawful essential properties of the things in question. A given system com-posed of parts of certain kinds necessarily comes with all its systemic prop-

---

[11] You do not make friends in mainstream analytic philosophy if you claim that possible-worlds semantics and metaphysics as well as modal logics are rather useless tools for a science-oriented philosophy.

[12] This is of course Putnam's example. The Kripkeans would disagree.

erties, where the necessity in question is *real* or *nomic*, not logical. So if consciousness or the mental in general is a systemic property of neuronal systems of a certain kind (or, if preferred, of whole brains), these systems will always and necessarily come with their systemic properties — under the given conditions. Accordingly, it makes no sense to ask why there is consciousness as though a certain system of a certain kind had a choice to not come with all its lawful properties. The existence of systemic properties is first of all a (ubiquitous) fact of nature, and an explanation of consciousness can only consist in the correct description of the special states or changes of state (mechanisms) of the brain that consist in being conscious. Asking for an explanation for why our brains come with consciousness at all rather than being zombie brains is like asking why there is something rather than nothing: it is a pseudo-question. For all these reasons, the zombie argument dissolves in the light of Bunge´s ontology or in the light of nomological essentialism, respectively (see, e.g., Garrett 2009, for a more detailed criticism).

Of course, one could argue that mental properties are different from other emergent properties because they are subjective, that is, one needs to be that neuronal system in a brain in a body in a certain state to "detect" or "access" (or rather have) mental properties. However, considering the ubiquity of systemic properties and of lawfulness throughout the entire world, we have good reasons to assume that mental properties are no different from other systemic properties. While it is true that mental properties are not physical properties (in the sense of physics), they are still material properties of material things. But then Bunge´s metaphysics combines substance monism with property pluralism: there are physical, chemical, biotic, social, and mental properties. However, the latter are not separated by an ontological gap from the rest of the world, as Chalmers seems to assume in his conception of property dualism: they just are yet another example of qualitative novelty in this world. This, at least, should in my view be the metaphysical working assumption of a (neuro)scientific approach to the mental.

It goes without saying that the implications of Bunge's metaphysics for the philosophy of mind deserve a much more extensive study than what I could do in this short overview here. So it must suffice to point out that by adopting a science-oriented and materialist metaphysics such as Bunge's, the philosophy of mind could avoid many fruitless debates and thus focus on important problems rather than self-inflicted pseudo-problems based on fragmentary metaphysics.

References

Bunge, Mario. 1967. *Scientific Research II: The Search for Truth*. Berlin: Springer-Verlag.

——. 1977. *Treatise on Basic Philosophy. Ontology I: The Furniture of the World*. Dordrecht: Reidel.

——. 1979. *Treatise on Basic Philosophy. Ontology II: A World of Systems*. Dordrecht: Reidel.

——. 1980. *The Mind-Body Problem*. Oxford: Pergamon Press.

——. 1983. *Treatise on Basic Philosophy*, vol. 6. *Epistemology & Methodology II: Understanding the World*. Dordrecht: Reidel.

Bunge, Mario & Rubén Ardila. 1987. *Philosophy of Psychology*. New York, NY: Springer-Verlag.

Bunge, Mario & Martin Mahner. 2004. *Über die Natur der Dinge. Materialismus und Wissenschaft* [in German]. Stuttgart: Hirzel-Verlag.

Chalmers, David J. 1996. *The Conscious Mind: In Search of a Fundamental Theory*. Oxford: Oxford University Press.

Ellis, Brian D. 2002. *The Philosophy of Nature: A Guide to the New Essentialism*. Chesham: Acumen.

Garrett, Brian J. 2009. Causal essentialism versus the zombie worlds. *Canadian Journal of Philosophy* 39(1): 93–112.

Glennan, Stuart. 2002. Rethinking mechanistic explanation. *Philosophy of Science* 69(S3): S342–S353.

Kary, Michael & Martin Mahner. 2002. How would you know if you synthesized a thinking thing? *Minds and Machines* 12(1): 61–86.

Kim, Jaegwon. 1978. Supervenience and nomological incommensurables. *American Philosophical Quarterly* 15(2): 149–156.

——. 1993. *Supervenience and Mind*. Cambridge: Cambridge University Press.

Machamer, Peter, Lindley Darden & Carl F. Craver. 2000. Thinking about mechanisms. *Philosophy of Science* 67(1): 1–25.

Mahner, Martin & Mario Bunge. 1997. *Foundations of Biophilosophy*. Berlin-Heidelberg: Springer-Verlag.

——. 2001. Function and functionalism: A synthetic perspective. *Philosophy of Science* 68(1): 75–94.

Miller, Steven M. 2007. On the correlation/constitution distinction problem (and other hard problems) in the scientific study of consciousness. *Acta Neuropsychiatrica* 19(3): 159–176.

Robb, David & John Heil. 2013. Mental causation. In Zalta, Edward N., ed, *The Stanford Encyclopedia of Philosophy*.

REFERENCES

Adler-Lomnitz, Larissa. 1975. *Cómo sobreviven los marginados* [in Spanish]. México, D.F.: Siglo xxi.

Agassi, Joseph. 1985. *Technology: Philosophical and Social Aspects*. Dordrecht: Kluwer.

Albert, Hans. 1988. Hermeneutics and economics. A criticism of hermeneutical thinking in the social sciences. *Kyklos* 41: 573–602.

Ampère, André-Marie. 1843. *Essai sur la philosophie des sciences* [in French], 2 vols. Paris: Bachelier.

Ayala, Francisco J. 2016. *Evolution, Explanation, Ethics, and Aesthetics: Towards a Philosophy of Biology*. Cambridge, MA: Elsevier.

Barber, Bernard. 1952. *Science and the Social Order*. London: George Allen & Unwin.

Barraclough, Geoffrey. 1979. *Main Trends in History*. New York, NY-London: Holmes & Meier.

Berkeley, George. 1901 [1710]. *A Treatise concerning the Principles of Human Knowledge*. In *Works*, vol. 1. Oxford: Clarendon Press.

Bertalanffy, Ludwig von. 1950. An outline of general systems theory. *British Journal for the Philosophy of Science* 1: 139–164.

Blackett, Patrick M.S. 1949. *Fear, War, and the Bomb*. New York, NY: Whittlesey.

Bohannon, John. 2015. Many psychology papers fail replication tests. *Science* 349: 910–911.

Boly, Melanie, Anil K. Seth, Melanie Wilke, Paul Ingmundson, Bernard Baars, Steven Laureys, David B. Edelman & Naotsugu Tsuchiya. 2013. Consciousness in human and non-human animals: new advances and future directions. *Frontiers in Psychology* 4: 1–20.

Boulding, Kenneth. 1956. General systems theory — the skeleton of science. *Management Science* 2: 197–208.

Braudel, Fernand. 1996 [1949]. *La Méditerranée et le monde méditerranéen à l'époque de Philippe II* [in French]. Berkeley, CA: University of California Press.

——. 1982 [1979]. *Civilization and Capitalism*, vol. 1: *The Wheels of Commerce*. New York, NY: Harper & Row.

Bruera, José Juan. 1945. *La lógica, el Derecho y la escuela de Viena* [in Spanish]. *Minerva* 2: 170–177.

Bunge, Mario. 1944. *Auge y fracaso de la filosofía de la naturaleza* [in Spanish]. *Minerva* 1: 213–235.

——. 1951. La fenomenología y la ciencia [in Spanish]. *Cuadernos Americanos* no. 4: 108–122. Revised version in *Una filosofía realista para el nuevo milenio*, pp. 265–285. Lima: Universidad Garcilaso de la Vega.

——. 1954. New dialogues between Hylas and Philonous. *Philosophy and Phenomenological Research* 15: 192–199.

——. 1955. The philosophy of the space–time approach to the quantum theory. *Methodos* 7: 295–308.

——. 1956. A survey of the interpretations of quantum mechanics. *American Journal of Physics* 24: 272–286.

——. 1957. *Filosofar científicamente y encarar la ciencia filosóficamente* [in Spanish]. *Ciencia e Investigación* 13: 244–254.

——. 1959a. *Metascientific Queries*. Evanston, IL: Charles C. Thomas.

——. 1959b. *Causality: The Place of the Causal Principle in Modern Science*. Cambridge, MA: Harvard University Press.

——. 1959c. Review of K. Popper's *The Logic of Scientific Discovery*. *Ciencia e investigación* 15: 216.

——. 1962a. An analysis of value. *Mathematicae Notae* 18: 95–108.

——. 1962b. Cosmology and magic. *The Monist* 44: 116–141.

——. 1966. Mach's critique of Newtonian mechanics. *American Journal of Physics* 34: 585–596.

——. 1967a. *Foundations of Physics*. Berlin-Heidelberg-New York, NY: Springer-Verlag.

——. 1967b. *Scientific Research*, 2 vols. Berlin-Heidelberg-New York, NY: Springer-Verlag. Rev. ed.: *Philosophy of Science*. New Brunswick, NJ: Transaction, 1998c.

——. 1967c. A ghost-free axiomatization of quantum mechanics. In M. Bunge, ed., *Quantum Theory and Reality*, pp. 105–117. Berlin-Heidelberg-New York, NY: Springer-Verlag.

——. 1967d. Physical axiomatics. *Reviews of Modern Physics* 39: 463–474.

——. 1967e. The structure and content of a physical theory. In M. Bunge, ed., *Delaware Seminar in the Foundations of Physics*, pp. 15–27. Berlin-Heidelberg,- New York, NY: Springer-Verlag.

——. 1967f. Analogy in quantum mechanics: From insight to nonsense. *British Journal for the Philosophy of Science* 18: 265–286.

——. 1968. On Mach's nonconcept of mass. *American Journal of Physics* 36: 167.

——. 1970. The so-called fourth indeterminacy relation. *Canadian Journal of Physics* 48: 1410–1411.

——. 1973a. *Philosophy of Physics*. Dordrecht: Reidel.

——. 1973b. *Method, Model and Matter*. Dordrecht: Reidel.

——. 1974a. *Treatise on Basic Philosophy*, vol. 1: *Sense and Reference*. Dordrecht: Reidel.

——. 1974b. *Treatise on Basic Philosophy*, vol. 2: *Interpretation and Truth*. Dordrecht: Reidel.

——. 1975. ¿Hay proposiciones? *Aspectos de la Filosofía de W. V. Quine* [in Spanish], pp. 53–68. Valencia: Teorema.

——. 1976. Review of Wolfgang Stegmüller's *The Structure and Dynamics of Theories*. *Mathematical Reviews* 55: 333, no. 2480.

——. 1977. *Treatise on Basic Philosophy*, vol. 3: *The Furniture of the World*. Dordrecht: Reidel.

——. 1979a. The Einstein–Bohr debate over quantum mechanics: Who was right about what? *Lecture Notes in Physics* 100: 204–219.

——. 1979b. *Treatise on Basic Philosophy*, vol. 4: *A World of Systems*. Dordrecht: Reidel.

——. 1980. *The Mind–Body Problem*. Oxford: Pergamon.

——. 1983a. *Treatise on Basic Philosophy*, vol. 5: *Exploring the World*. Dordrecht: Reidel.

——. 1983b. *Treatise on Basic Philosophy*, vol. 6: *Understanding the World*. Dordrecht: Reidel.

——. 1985a. *Treatise on Basic Philosophy*, vol. 7, Part I: *Philosophy of Science: Formal and Physical Sciences*. Dordrecht: Reidel.

——. 1985b. *Treatise on Basic Philosophy*, vol. 7, Part II: *Philosophy of Science: Life Science, Social Science and Technology*. Dordrecht: Reidel.

——. 1987. *Philosophy of Psychology*. Berlin-Heidelberg-New York, NY: Springer-Verlag.

——. 1989. *Treatise on Basic Philosophy*, vol. 8: Ethics: *The Good and the Right*. Dordrecht-Boston: Reidel.

——. 1991/92. A critical examination of the new sociology science. *Philosophy of the Social Sciences* 21: 524–560; 22: 46–76.

——. 1992. Sette paradigmi cosmologici: L'animale, la scala, il fiume, la nuvola, la macchina, il libro e il sistema dei sistemi [in Italian]. *Aquinas* 35: 219–235.

——. 1994. L'écart entre les mathématiques et le réel. In M. Porte, Ed., *Passion des formes* [Festchrift for René Thom] vol. 1, pp. 165–173. Fontenay-St. Cloud: E.N.S Editions.

——. 1996. *Finding Philosophy in Social Science*. New Haven, CT: Yale University Press.

——. 1997. Moderate mathematical fictionism. In Agazzi, Evandro & György Darwas, eds. *Philosophy of Mathematics Today*, pp. 51–71. Dordrecht-Boston: Kluwer Academic.

——. 1998. *Social Science Under Debate*. Toronto: University of Toronto Press.

——. 1999a. *The Sociology–Philosophy Connection*. New Brunswick, NJ: Transaction.

——. 1999b. *Philosophical Dictionary*, enlarged ed. Amherst, NY: Prometheus Books.

——. 2001. *Philosophy in Crisis: The Need for Reconstruction*. Amherst, NY: Prometheus Books.

——. 2003a. *Emergence and Convergence*. Toronto: University of Toronto Press.

——. 2003b. Velocity operators and time–energy relations in relativistic quantum mechanics. *International Journal of Theoretical Physics* 42: 135–142.

——. 2006. *Chasing Reality: The Strife over Realism*. Toronto: University of Toronto Press.

——. 2007. Did Weber practice the philosophy he preached? In McFalls, Laurence, ed. *Max Weber's "Objectivity" Revisited*, pp. 119–134. Toronto: University of Toronto Press.

——. 2008. Bayesianism: Science or pseudoscience? *International Review of Victimology* 15: 169–182. Repr. In Bunge 2012a.

——. 2009a. The failed theory behind the 2008 crisis. In Mohamed, Cherkaoui & Peter Hamilton, eds., *Raymond Boudon: A Life in Sociology*, vol. 1, pp. 127–142. Oxford: Bardwell.

——. 2009b. Advantages and limits of naturalism. In Shook, John R. & Paul Kurtz, eds., *The Future of Naturalism*. Amherst, NY: Humanities Books.

——. 2009c. *Political Philosophy: Fact, Fiction, and Vision*. New Brunswick, NJ: Transaction Publishers.

——. 2010. *Matter and Mind*. Boston Studies in the Philosophy of Science, vol. 287. New York, NY: Springer.

——. 2011. Knowledge: Genuine and bogus. *Science & Education* 20: 411–438.

——. 2012a. *Evaluating Philosophies*. Boston Studies in the Philosophy of Science, vol. 295. New York, NY: Springer.

——. 2012b. The correspondence theory of truth. *Semiotica* 188: 65–76.

——. 2013. *Medical Philosophy: Conceptual Issues in Medicine*. Singapore: World Scientific.

——. 2014. Wealth and wellbeing, economic growth, and integral development. *International Journal of Health Services* 44: 835–844.

——. 2015. Does the Aharonov–Bohm effect occur? *Foundations of Science* 20: 129–133.

——. 2016a. *Between Two Worlds: Memoirs of a Philosopher Scientist*. Switzerland: Springer International.

——. 2016b. Why axiomatize? *Foundations of Science*, in press.

Buss, David M. 2015. *Evolutionary Psychology: The New Science of the Mind*, 5th ed. New York, NY: Routledge.

Carnap, Rudolf. 1928. *Der logische Aufbau der Welt*. Transl.: *The Logical Structure of the World. Pseudoproblems in Philosophy*. Berkeley, CA: University of California Press.

Carnap, Rudolf. 1936. Testability and meaning. *Philosophy of Science* 4: 419–471.

Cirelli, Marco. 2015. Status of (direct and) indirect dark matter searches. *Proceedings of Science* arXiv: 1511.02031v3 [astro-ph.HE].

Condorcet, Nicholas. 1976. *Selected Writings*. K. M. Baker, ed. Indianapolis, IN: Bobbs-Merrill.

Costanza, Robert *et al.* 2014. Time to leave GDP behind. *Nature* 505: 283–285.

Covarrubias, Guillermo M. 1993 An axiomatization of general relativity. *International Journal of Theoretical Physics* 32: 1235–1254.

Cravioto, Joaquin. 1958. Protein metabolism in chronic infantile malnutrition (kwashiorkor). *American Journal of Clinical Nutrition*. 6: 495–503.

d'Holbach, Paul-Henri Thiry. 1770. *Système de la Nature ou Des Loix du Monde Physique et du Monde Moral* [in French], 2 vols. London: M.-M. Rey.

——. 1994 [1773]. *Système Social, ou Principes Naturels de la Morale et de la Politique*. Paris: Fayard.

Daston, Lorraine & Peter Galison. 2007. *Objectivity*. New York, NY: Zone Books.

Dawkins, Richard. 1976. *The Selfish Gene*. Oxford: Oxford University Press.

de Solla Price, Derek. 1963. *Little Science, Big Science*. New York, NY: Columbia University Press.

Descartes, René. 1974 [1664]. *Oeuvres* [in French], vol. XI, Charles Adam and Paul Tannéry, eds. Paris: Vrin.

Dilthey, Wilhelm. 1959 [1883]. *Einleitung in die Geisteswissenschaften*. In *Gesammelte Schriften* [in German], vol. 1. Stuttgart: Teubner; Göttingen: Vanderhoeck und Ruprecht.

Dirac, Paul A.M. 1958. Generalized Hamiltonian dynamics. *Proceedings of the Royal Society of London* 246: 326–332.

Duhem, Pierre. 1908. SWZEIN TA FAINOMENA: *Essai sur la théorie physique de Platon à Galilée* [in French]. Paris: Hermann.

Durkheim, Emile. 1988 [1895]. *Les règles de la méthode sociologique* [in French]. Paris: Flammarion.

Dyzenhaus, David. 1997. Legal theory in the collapse of Weimar. Contemporary lessons? *American Political Science Review* 91: 121–134.

Einstein, Albert. 1950. *Out of my Later Years*. New York, NY: Philosophical Library.

Einstein, Albert, Boris Podolsky & Nathan Rosen. 1935. Can quantum-mechanical description of physical reality be considered complete? *Physical Review* 47: 777–789.

Ellis, George & Joe Silk. 2014. Defend the integrity of physics. *Nature* 516: 321–323.

Engels, Frederick. 1941. *Dialectics of Nature*. London: Lawrence & Wishart.

Everett, Hugh. *1957*. Relative state formulation of quantum mechanics. *Reviews of Modern Physics* 29: 454–462.

Falk, Gottfried & Herbert Jung. 1959. Axiomatik der Thermodynamik. *Handbuch der Physik* [Encyclopedia of Physics], pp. 119–175. Berlin-Heidelberg: Springer-Verlag

Feuerbach, Ludwig. 1947. *Kleine philosophische Schriften (1842–1845)* [in German]. Leipzig: Verlag Felix Meiner.

Feynman, Richard P. 1949. Space–time approach to quantum electrodynamics. *Physical Review* 76: 769–789.

Fogel, Robert W. 1994. Economic growth, population theory, and physiology: The bearing of long-term processes on the making of economic policy. *American Economic Review* 84: 369–395.

Fontana, Josep. 2011. *Por el bien del imperio: una historia del mundo desde 1945* [in Spanish]. Barcelona: Pasado y Presente.

Foster, Jacob G., Andrey Rzhetsky & James A. Evans. 2015. Tradition and innovation in scientists' research strategies. *American Sociological Review* 80: 875–908.

Fraassen, Bas C. van 1980. *The Scientific Image*. Oxford: Clarendon Press.

Fuller, Leon L. 1958. Positivism and fidelity to law: A reply to Professor Hart. *Harvard Law Review* 91: 121–134.

Galilei, Galileo. 1953 [1693]. *Il saggiatore* [in Italian]. In *Opere*. Milano, Napoli: Riccardo Ricciardi.

Galison, Peter. 1987. *How Experiments End*. Chicago, IL: University of Chicago Press.

Garcia, John. 1981. Tilting at the windmills of academia. *American Psychologist* 36: 149–158.

Geertz, Clifford. 1973. *The Interpretation of Cultures*. New York, NY: Basic Books.

Gintis, Herbert, Samuel Bowles, Robert Boyle & Ernst Fehr, eds. 2005. *Moral Sentiments and Material Interests: The Foundations of Cooperation in Economic Life*. Cambridge, MA: MIT Press.

Gordin, Michael D. 2015. Myth 27. That a clear line of demarcation has separated science from pseudoscience. In Numbers, Ronald L. & Kostas Kampourakis, eds. *Newton's Apple and Other Myths About Science*, pp. 219–226. Cambridge, MA: Harvard University Press.

Gould, Stephen Jay. 1990. *The Panda's Thumb*. London: Penguin.

Gruber, Howard E. & Paul H. Barrett. 1974. *Darwin on Man*. New York, NY: E.P. Dutton.

Hacohen, Malachi Haim. 2000. *Karl Popper: The Formative Years 1902–1945*. Cambridge: Cambridge University Press.

Halmos, Paul. 1960. *Naïve Set Theory*. New York, NY: Van Nostrand Reinhold.

Harris, Marvin. 1968. *The Rise of Anthropological Theory*. New York, NY: Crowell.

Hart, H. L. A. 1961. *The Concept of Law*. Oxford: Oxford University Press.

Hastings, Max. 2015. *The Secret War: Spies, Codes and Guerrillas 1939–1945*. London: Wílliam Collins.

Hayek, Friedrich von. 1952. *The Counter-Revolution of Science*. Glencoe, IL: Free Press.

Hebb, Donald O. 1949. *The Organization of Behavior: A Neuropsychological Theory*. New York, NY: Wiley & Sons.

——. 1951. The role of neurological ideas in psychology. *Journal of Personality* 20: 39–55.

——. 1980. *Essay on Mind*. Hillsdale, NJ: Erlbaum.

Heidegger, Martin. 1993 [1926]. *Sein und Zeit* [in German]. Tübingen: Max Niemeyer.

Heisenberg, Werner. 1969. *Der Teil und das Ganze* [in German]. Munich: Piper.

Henry, Richard Conn. 2005. The mental universe. *Nature* 436: 29.

Hilbert, David. 1918. Axiomatisches Denken [in German]. *Mathematische Annalen* 78: 405–415.

——. 1935. *Gesammelte Abhandlungen*, Vol. 3. Berlin: Julius Springer.

Hume, David. 1902 [1748]. *An Enquiry Concerning Human Understanding*, 2nd ed. Oxford: Clarendon Press.

——. 2000 [1739]. *A Treatise on Human Nature*, new ed. Oxford: Oxford University Press.

Husserl, Edmund. 1931 [1913]. *Ideas: General Introduction to Pure Phenomenology*. London: George Allen & Unwin.

——. 1995 [1928]. *Cartesianische Meditationen* [in German]. Hamburg: Meiner.

——. 1970 [1935]. *The Crisis of European Sciences*. Evanston, IL: Northwestern University Press.

Ingenieros, José. 1917. *Hacia una moral sin dogmas* [in Spanish]. Buenos Aires: Rosso.

Ioannidis, John P.A. 2005. Why most published research findings are false. *PLoSMed* 2(8): e124.

Israel, Jonathan. 2010. *A Revolution of the Mind*. Princeton, NJ: Princeton University Press.

——. 2014. *Revolutionary Ideas*. Princeton, NJ: Princeton University Press.

Kahneman, Daniel. 2011. *Thinking, Fast and Slow*. New York, NY: Farrar, Strauss and Giroux.

Kant, Immanuel. 1952 [1787]. *Kritik der reinen Venunft* [in German], 2nd ed. Hamburg: Felix Meiner.

——. 1912. *Briefwechsel* [in German], 3 vols. Munich: Georg Muller.

Kelsen, Hans. 1945. *General Theory of Law and State*. Cambridge, MA: Harvard University Press.

Koepsell, David. 2009. *Who Owns You? The Corporate Rush to Patent Your Genes*. Malden, MA: Wiley-Blackwell.

Kuhn, Thomas S. 1977. *The Essential Tension: Selected Studies in Scientific Tradition and Change*. Chicago, IL: University of Chicago Press.

Lalande, André. 1938. *Vocabulaire technique et critique de la philosophie* [in French], 2nd ed., 3 vols. Paris: Alcan.

Lange, Friedrich Albert. 1905 [1875]. *Geschichte des Materialismus* [in German], 2nd ed., 2 vols. Leipzig: Philipp Reclam.

Latour, Bruno. 1987. *Science in Action: How to Follow Scientists and Engineers Through Society*. Cambridge, MA: Harvard University Press.

Latour, Bruno & Steven Woolgar. 1979. *Laboratory Life. The Construction of Scientific Facts*. Princeton, NJ: Princeton University Press.

Laudan, Larry. 1988. The demise of the demarcation problem. In Ruse, Michael, ed. *But Is It Science?: The Philosophical Question in the Creation/Evolution Controversy*, pp. 337–350. Amherst, NY: Prometheus Books.

Le Dantec, Félix. 1912. *Contre la métaphysique* [in French]. Paris: Alcan.

Ledford, Heidi. 2015. Team science. *Nature* 525: 308–311.

Leibniz, Gottfried Friedrich. 1981 [1703]. *New Essays on Human Understanding*. Cambridge: Cambridge University Press.

Lenin, V[ladimir] I[lich]. 1981. *Collected Works*, vol. 38: *Philosophical Notebooks* [1914–15]. Moscow: Foreign Languages Publishing House.

Lewontin, Richard. 2000. *It Ain't Necessarily So*. New York, NY: New York Review Books.

Locke, John. 1975 [1690]. *An Essay on the Human Understanding*. Oxford: Oxford University Press.

Lundstedt, Anders V. 1956. *Legal Thinking Revised: My Views on Law*. Stockholm: Almqvist & Wiksell.

Mach, Ernst. 1893 [1883]. *The Science of Mechanics*. La Salle, IL: Open Court.

Mahner, Martin, ed. 2001. *Scientific Realism: Selected Essays of Mario Bunge*. Amherst, NY: Prometheus Books.

Mahner, Martin & Mario Bunge. 1997. *Foundations of Biophilosophy*. New York, NY: Springer.

McKinsey, John C.C., A.C. Sugar & Patrick Suppes. 1953. Axiomatic foundations of classical particle mechanics. *Journal of Rational Mechanics and Analysis* 2: 253–272.

Merton, Robert K. 1973. *Sociology of Science*. Chicago, IL: University of Chicago Press.

Merton, Robert K. & Elinor Barber. 2004. *The Travels and Adventures of Serendipity*. Princeton, NJ: Princeton and Virginia Press.

Meyerson, Émile. 1931. *Du cheminement de la pensée* [in French], 3 vols. Paris: Alcan.

Mill, John Stuart. 1952 [1843]. *A System of Logic Ratiocinative and Inductive*. London: Longmans, Green.

Mirowski, Philip. 2011. *Science-Mart: Privatizing American Science*. Cambridge, MA: Harvard University Press.

Moulines, Carlos Ulises. 1975. A logical reconstruction of simple equilibrium thermodynamics. *Erkenntnis* 9: 101–130.

——. 1977. Por qué no soy materialista [in Spanish]. *Crítica* 9: 25–37.

Natorp, Paul. 1910. *Die logische Grundlagen der een Wissenschafteni* [in German]. Leipzig-Berlin: B.G. Teubner.

Neuberger, David. 2016. Stop the needless dispute of science in the courts. *Nature News*. Retrieved from http://www.nature.com/news/stop-needless-dispute-of-science-in-the-courts-1.19466

Neurath, Otto. 1955. Encyclopedia and unified science. In Otto Neurath, Rudolf Carnap & Chales Morris, eds. *International Encyclopedia of Unified Science*, vol. I, no. 1. Chicago, IL: University of Chicago Press.

Newton, Isaac. 1999 [1687]. *The Principia: Mathematical Principles of Natural Philosophy*. Berkeley, CA: University of California Press.

Numbers, Ronald L. & Kostas Kampourakis, eds. 2015. *Newton's Apple and Other Myths About Science*. Cambridge, MA: Harvard University Press.

Omnès, Roland. 1999. *Understanding Quantum Mechanics*. Princeton, NJ: Princeton University Press.

Open Science Collaboration [a group of 270 psychologists from around the world]. 2015. Estimating the reproducibility of psychological science. *Science* 349: 943.

Owens, Brian. 2016. Access all areas. *Nature* 533: S71–S72.

Pavlov, Ivan. 1960 [1927]. *Conditioned Reflexes*. New York, NY: Dover.

Peng, Yueqing, Sarah Gillis-Smith, Hao Jin, Dimitri Tränkner, Nicholas J.P. Ryba & Charles S. Zuker. 2015. Sweet and bitter taste in the brain of awake behaving animals. *Nature* 527: 512–15.

Pérez-Bergliaffa, Santiago. 1997. Toward an axiomatic pregeometry of space–time. *International Journal of Theoretical Physics* 37: 2281–2299.

Pérez-Bergliaffa, Santiago, Gustavo Romero & Héctor Vucetich. 1993. Axiomatic foundation of non-relativistic quantum mechanics: A realist approach. *International Journal of Theoretical Physics* 32: 1507–1525.

Pievani, Telmo. 2014. *Evoluti e abandonati* [in Italian]. Torino: Einaudi. Ponsa.

Pinker, Steven. 2002. *The Blank Slate: The Modern Denial of Human Nature*. New York, NY: Penguin.

Popper, Karl R. 1959 [1935]. *The Logic of Scientific Discovery*. London: Hutchinson.
——. 1960. *The Poverty of Historicism*, 2nd ed. London: Routledge & Kegan Paul.

Popper, Karl R. & John C. Eccles. 1977. *The Self and Its Brain*. Heidelberg-New York, NY: Springer-Verlag.

Pound, Roscoe. 1931. The call for a realist jurisprudence. *Harvard Law Review* 44: 697–711.

Press, William H. 2013. What's so special about science (and how much should we spend on it)? *Science* 342: 817–822.

Puccini, Gabriel D., Santiago Pérez-Bergliaffa & Héctor Vucetich. 2008. Axiomatic foundations of thermostatics. *Nuovo Cimento B*. 117: 155–177.

Quine, Willard Van Orman. 1969. Epistemology naturalized. In Willard Quine, ed. *Ontological Relativity and Other Essays*, pp. 69–90. New York, NY: Columbia University Press.

Quine, Willard Van Orman & Nelson Goodman. 1940. Elimination of extralogical predicates. *Journal of Symbolic Logic* 5: 104–109.

Quintanilla, Miguel A. 2005. *Tecnología: Un enfoque filosófico*. Mexico D.F.: Fondo de Cultura Económica.

Raynaud, Dominique. 2015. *Scientific Controversies*. New Brunswick, NJ: Transaction.

———. 2016. *Qu'est ce que la technologie?* [in French]. Paris: Editions Matériologiques.

Renan, Ernest. 1949 [1852]. *Avérroès et l'avérrroisme* [in French]. *Oeuvres Complètes*, vol. III. Paris: Calmann-Lévy.

Rennie, Drummond. 2016. Make peer review scientific. *Nature* 535: 31–33.

Rescher, Nicholas. 1985. *The Strife of Systems*. Pittsburgh, PA: University of Pittsburgh Press.

Ridley, Matt. 2016. *The Evolution of Everything: How New Ideas Emerge*. New York, NY: Harper.

Romero, Gustavo E. & Gabriela S. Vila. 2014. *Introduction to Black Hole Astrophysics. Lecture Notes in Physics*, vol. 876. Berlin, Heidelberg: Springer-Verlag.

Rousseau, Jean-Jacques. 2009 [1762]. *Émile, ou De l'éducation* [in French]. Paris: Flammarion.

Russell, Bertrand. 1954 [1927]. *The Analysis of Matter*. New York, NY: Dover.

———. 1995 [1940]. *An Inquiry into Meaning and Truth*. London: Routledge.

Schlosshauer, Maximilian. 2007. *Decoherence and the Quantum-to-Classical Transition*. Berlin-Heidelberg-New York, NY: Springer-Verlag.

Schöttler, Peter. 2013. *Scientisme: Sur l'histoire d'un concept difficile* [in French]. *Revue de Synthèse* 134: 89–113.

———. 2015. *Die "Annales" — Historiker und die deutsche Geschichtswissenschaft*. Tübingen: Mohr-Siebeck.

Smolin, Lee. 2006. *The Trouble with Physics*. New York, NY: Penguin.

Sneed, Joseph D. 1971. *The Logical Structure of Mathematical Physics*. Dordrecht: Reidel.

Sokal, Alan & Jean Bricmont. 1997. *Fashionable Nonsense*. New York, NY: Picador.

Sperber, Jonathan. 2013. *Karl Marx: A Nineteenth-Century Life*. New York, NY-London: Liveright.

Stone, Julius. 1966. *Social Dimensions of Law and Justice*. Stanford, CA: Stamford University Press.

Stove, David. 1982. *Popper and After: Four Modern Irrationalists*. Oxford: Pergamon Press.

Szyf, Moshe, Patrick McGowan & Michael J. Meaney. 2008. The social environment and the epigenome. *Environmental and Molecular Mutagenesis* 49: 46–60.

Takahashi, Daniel Y., Alicia Fenley, Yayoi Teramoto & Asif A. Ghazanfar. 2015. The developmental dynamics of marmoset monkey vocal production. *Science* 349: 734–738.

Tarski, Alfred. 1944. The semantical concept of truth and the foundations of semantics. *Philosophy and Phenomenological Research* 4: 341–375.

Tola, Fernando & Carmen Dragonetti. 2008. *Filosofía de la India* [in Spanish]. Barcelona: Kairós.

Trigger, Bruce G. 2003. *Artifacts & Ideas*. New Brunswick, NJ: Transaction Publishers.

Truesdell, Clifford. 1984. *An Idiot's Fugitive Essays on Science*. New York, NY: Springer-Verlag.

Wang, Yu-Ze Poe. 2011. *Reframing the Social*. Surrey: Ashgate.

Weber, Max. 1924. *Die sozialen Gründe des Untergangs der antiken Kultur* [in German]. In von Clemens, Bauer, ed. *Gesammelte Aufsätze zur Wirtschafts-und Sozialgeschichte*, pp. 289–311. Tübingen: Mohr.

——. 1976 [1921]. *Wirtschaft und Gesellschaft: Grundriss derverstehende Soziologie* [in German], 3 vols. Tübingen: Mohr.

——. 1988a [1904]. *Die "Objektivität" sozialwissenschaftlicher und sozialpolitiker Erkenntnis* [in German]. In von Clemens, Bauer, ed. *Gesammelte Aufsätze zur Wissenschaftslehre*, pp. 146–214. Tübingen: Mohr.

——. 1988b [1913]. *Ueber einige Kategorien der verstehende Soziologie* [in German]. In *Gesammelte Aufsätze zur Wissenschaftslehre*, pp. 427–474. Tübingen: Mohr.

Wikström, Per-Olof & Robert J. Sampson, eds., *The Explanation of Crime*. Cambridge: Cambridge University Press.

Wilsdon, James. 2015. We need a measured approach to metrics. *Nature* 523: 129.

Worden, Frederic G., Judith P. Swazey & George Adelman, eds. 1975. *The Neurosciences: Paths of Discovery*. Cambridge, MA: MIT; New York, NY: Straus, Farrar and Giroux.

Zeilinger, Anton. 2010. *Dance of the Photons: From Einstein to Quantum Teleportation*. New York, NY: Farrar, Straus and Giroux.

Zuckerman, Harriet. 1977. *Scientific Elite: Nobel Laureates in the United States*. New York, NY: Free Press.

INDEX

determinism, vi, 163, 166, 168, 170–71
  genetic 23, 47, 159
diabetes mellitus, 95
diagnostic problems, 4
dialectics, 54, 126, 158
Diderot, Denis, 30, 125, 138, 152
Dieudonné, Jean, 19
Dilthey, Wilhelm, 58, 109, 118, 126–27, 140–42, 158, 204
Dirac, Paul A.M., 13, 31, 70–71, 76, 204
direct problems, 4, 99
direct/inversedistinction, 4, 11, 99. *See also under* problems
discoveries, 36–37, 42, 107, 156
Discovery Voyages, 116
disinterestedness, 56
DNA testing, 145
Dragonetti, Carmen, 107, 210
dual axiomatics, 65–68, 74–77
dualism, 43, 163, 195
  psychoneural, xiii, 40, 126
Duhem, Pierre, 26, 56, 108, 139, 204
Durkheim, Emile, 126, 137, 141, 204
Dyzenhaus, David, 115, 204

$E = mc^2$, 97, 119
Eccles, John C., 43, 208
economic rationality, 43
economism, 131
egology, 28, 106. *See also* Husserl; phenomenology
Einstein, Albert, 29, 32, 37, 40–41, 51–52, 54, 56, 62, 69, 97, 120, 138, 204. *See also* general relativity; special relativity
electrodynamics, xiv, 11, 12, 54
  classical, 51, 68, 76, 97, 152

Ellis, Brian D., 180, 181n, 196
Ellis, George, 13, 204
Ellul, Jacques, 154
Empedocles, 9, 30
empiricism, v, 35, 45, 95, 100
empiricists, xv, 4, 59, 93, 97, 107, 109
Empiricus, Sextus, 107
endoheresy, 54
endostructure, 188
energetism, 118–19
Engels, Frederick, 20, 21, 44, 152, 204
Enlightenment, 117, 139–40. *See also* French Enlightenment; Late Austrian Enlightenment; Scottish Enlightenment
entanglement, 120
environment, 21, 96
  of a system, 188–89
  negative effects on, 153
  protection of, 159
  sustainability of, 132
Eötvös, Loránd, 62
Epicureanism, 123
Epicurus, 29, 123
epigenetics, 22, 47, 145
epistemic communism, 56
epistemological phenomenalism, 26, 108, 187
epistemology, 98, 101, 116–17, 119, 123
épokhé, 106
equivalence relation, 81
Erathostenes, 9
Erlangen program, 19
Escher, Maurits, 86, 109
essential properties, 180, 182, 184, 194
essential singularity, 154

markers, v, 77, 95, 96
Marx, Karl, 20, 133, 151, 156, 158
Communist Manifesto, 21, 130
Marxists, 51, 54–55, 58, 126, 156
mass, 30, 62, 67, 97, 118–19, 178,
   181–82
material infrastructure, 151
material thing, 21, 92, 178, 184
materialism, ix, xiii, xiv, 123–136
   dialectical, 21, 126
   emergentist, 187, 193
   historical, 54–55, 131, 141
   scientific, xiv, 118, 136
   systemic or total, 134–136
   vulgar, 20
materiality, 81, 126
mathematical auxiliary, 66, 68
mathematical existence, 83
mathematical formalism, vi, 65, 67
mathematical model theory, 96
matter, 117–18, 128–29, 135–36, 192
Maupertuis, Pierre Louis Moreau de,
   30
Maxwell, James Clerk, 41, 51, 54, 97,
   150
McGowan, Patrick, 210
McKinsey, John C. C., 67, 207
Meaney, Michael J., 210
meaning, v, 100, 105
measure, v
measurement, v, 18, 39, 40, 57, 63,
   71, 77, 87, 95, 96, 97
   apparatus, 17, 63, 96
   general theories of, 95
mechanics, xiii, 51, 97, 190n9
mechanismic explanations, 190
mechanism, xiv, 19–20, 123
mechanisms, 190–91nn9–10
   of action, 45

Meeks, Thomas W., 172, 175
Mendeléyev, Dmitri, 39
mental, the, vi, 35–36, 72–73, 135–36
mentors, 1, 24, 46–47, 140, 156
Merton, Robert K., 11, 14, 16, 54,
   56–57, 143, 207
metaphysics, 116–17, 119–20, 177–78
   substance, 178, 195
methodology, 98, 100, 113
Meyerson, Émile, 56, 138, 207
microbiology, 10
military intelligence, 55
Mill, John Stuart, 110, 159, 207
Miller, Stanley, 10, 12, 177n, 189n,
   197
Milner, Peter, 36
mind, 38, 72–73, 119, 136, 177, 189
mind over matter, 25, 33, 128
Minkowski, Oscar, 95
Mirowski, Philip, 42, 207
models muddle, 61–63
modernity, 124, 131, 151, 157
molecular biology, 12, 21, 98
Moleschott, Jakob, 20
moral ambivalence, 154–55
Moulines, Carlos Ulises, 61, 62, 68,
   207
multiple realizability, 136, 192–93
multiverse, 87
Murakami, Masayoshi, 170, 175
mutability, 81
Myrdal, Gunnar, 143
mysterians, 144

Nagasaki, 55, 153
Nahmias, Eddy, 165, 169, 170, 172,
   173
nativists, xiii
Natorp, Paul, 110, 208

natural inequalities, 133
natural law, 115
natural/cultural dichotomy, 142
naturalism, 130, 131–34, 139
naturalized epistemology, 129
naturism, 132
Naturphilosophien, 29
Navon,David, 164, 175
Nazism, 113, 116, 154
negation, 20, 38, 119, 183
neo-Kantians, 97, 110, 125, 126
neoclassical economic theorists, 140
Neptune, 32, 101
Neuberger, David, 34, 208
neural systems, 73, 172
Neurath, Otto, 139–40, 208
neurons, 118, 136, 188, 193
neutrino, 37–38, 46
new essentialism, 180
Newton, Isaac, 9, 19, 28, 52–53, 120, 152, 208
    laws of motion, 62, 97, 118
    Principia, 67
Newtonian mechanics, 65, 117, 158
Nichols, Shaun, 173, 175
Nietzsche, Friedrich, 125, 146, 158
Nobel prizes, 36, 37, 46, 52, 56, 67, 73, 99, 118, 155, 156, 158
nomological essentialism, 180, 193, 195
nomological state space, 184–86, 188, 193–94
non-Euclidean geometrics, 101
non-triviality, 43
novelty, 19, 73, 195
nuclear bombs, 55, 155
Numbers, Ronald L., xi, 44, 205, 208

O'Keefe, John, 36, 73
objectivity, xii, 16, 97, 104, 109, 120, 126
obscurantist, 137
observables, 26, 71
Ockham, William of, 87
Oedipus, 40, 103
Olds, James, 36
Olympia Akademie, 52
Omnès, Roland, 69, 208
ontological commitment, 80
ontological nihilism, 79
ontological phenomenalism, 26, 107
ontology, vii, 81, 116–17, 119, 123, 128, 131, 178–79, 183, 186, 192, 195
Oparin, Aleksander, 12
Open Science Collaboration, 100, 208
operationalists, 62
operationism, 52
Operations Research, 145
Oppenheimer, J. Robert, 53, 153
organized skepticism, 56
originality, xii, 14, 45, 126
Ostwald, Willhelm, 118
Owens, Brian, 208
ownership, 43

Padoa, Alessandro, 65, 67, 81
paradox, 6, 21, 60, 107, 132
participant universe, 105
particle accelerators, xiv
Pasteur, Louis, 10
Paul, Saint, 133
Pauli, Wolfgang, 31, 37, 70
    theory of spin, 65
Pauling, Linus, 52, 98

Pavlov, Ivan, 2, 126, 208
Peano, Giuseppe, 65, 83–84
peer review, 42
Peirce, Charles Sanders, 11, 80, 159
Penfield, Wilder, 36
Peng, Yueqing, xiii, 208
Pepys, Samuel, 9, 52–53
Pérez-Bergliaffa, Santiago, 64, 121, 208, 209
periodic table, 39
Perrin, Jean, 29
peta, 91, 93, 96
pharmacology, 17, 150
phenomenal existence, 82, 88
phenomenalism, 26–28, 32, 52, 101, 106–8, 110, 138, 139
epistemological, 26, 108, 187
phenomenological reductions, 106
phenomenology, 28, 106
Philolaos, 28
philosopher-scientists, 159
philosophical matrix of scientific research, v-vi, 18, 143–46, 173
philosophical realism, 64, 69, 104, 120
philosophies of nature, 159, 180
philosophy of mind, vi-vii, ix, xiii, 177–78, 191–92
physicalism, xiii, 111, 134, 136
Pievani, Telmo, 22, 208
Pinker, Steven, xiii, 14, 208
Planet 9, 39
Plato, 86, 114, 123, 136
pluralism, 135, 195
Poincaré, Henri, 51, 55, 60, 97
Popper, Karl, v, 2, 12, 19, 35, 38–40, 43, 47, 100, 113, 120, 139
falsifiability of, 35, 38, 100
positive feedback, 147, 151

positivism, 28, 101, 110, 126, 137–38
legal, 115
logical, 2, 26, 60, 101, 163
possible-worlds fantasists, 83
postmodernists, 36, 51, 55–58, 125, 146
Pound, Roscoe, 115, 208
precision, xii, 13, 42, 76, 121
Press, William H., 147, 208
presuppositions, xi, 16–18, 25, 36, 49, 63, 75, 77, 192
primary properties, xiii, 26, 180
privatization, 41, 155
probabilities, 4, 6, 10, 34, 43, 70, 76, 100, 170, 171
problems, ix, 1–10, 16, 32–33
choice of, 1, 7, 46
cognitive, 147, 149
direct, 4, 9, 95, 99
inverse, 3–5, 9, 95, 99, 141
practical, 151
problem system, 11–12
technological, 149
types of, 2
prognosis, 4
properties, 178–85, 191–92
propositions, 6, 82, 89, 92
pseudoproblem, 5, 6, 159, 177
pseudoscience, 25, 86, 156–57, 159
conceptual matrix of, 157 fig 11.2
psychoanalysis, 23, 40, 159
psychoneural dualism, 40, 126
psychoneural theories, 73
Ptolemy, 25, 28, 108–9
Puccini, Gabriel D., 61, 64, 209
pulse (biological), 9, 48, 96
Putnam, Hilary, 119, 120, 194
Pythagoras, 8

subjectivism, xi, 28, 52, 71, 114, 128
subjectivists, 33, 60, 70, 113
substance, 178–79, 191–92, 195
success criterion, 35
Sugar, A. C., 207
Suppes, Patrick, 67–68, 207
Suppes-Sneed-Stegmüller structuralist
    school, 67
supranatural, 132–33
sustainability, 132
Swazey, Judith P., 210
symptoms, 4, 144
system, 4–5, 120, 147, 187–96,
systemic properties, 181, 185, 187,
    193, 195
systemism, ix, 123–36
Szyf, Moshe, 145, 210

Takahashi, Daniel Y., 24, 210
Tarski, Alfred, 61, 65, 65, 80, 113,
    210
Taylor, Charles, 141
technological designs, 152
technology, ix, 8, 14, 18, 32–33, 57,
    108, 113, 120, 149–156
technophilia, 153–54
technophobia, 153–54
technoscience, 152–53
tensor calculus, 52
teoritas, 96
testability, v, 13, 39,43, 100, 203
testimonies, 34
textualism, 108, 141
theodicy, 5
theoretical models, 61–62, 96–99
theory, 32, 34, 38, 48, 58, 59, 63, 65,
    97, 121 *See also* theoretical models
theory of mind, 140
theory-guided experiments, 98

thermodynamics, 29, 61, 62, 83, 166,
    207
things in themselves, 69, 110, 119, 125
thinking machines, 192–96
Thirty Years' War, 55, 116
Thom, René, 71
Thomas theorem, 105, 139
Thucydides, 112
time, 62, 70, 81, 92, 95, 117, 164n1,
    165, 179 fig A2.1, 184
Tola, Fernando, 107, 210
toxic gases, 154
Tränkner, Dimitri, 208
translations, 150, 154–55
Treaty of Versailles, 116
Trigger, Bruce G., 54, 143, 210
tropes, 183
Truesdell, Clifford, 45, 61, 68, 210
truths, v, xii, 2
    factual, 36, 92, 103–4, 113
    new, 14, 33, 35–36, 150
    objective, 146–47
    of fact, 113
tests, 48
truth maker, 92
truth-value, 3, 6, 38, 92, 100
Turing, Alan, 2
Turing machines, 72, 73

Ulam, Stanislav, 2
ultimatum game, 44
unconscious processes, 164–70
universality, 56
unseen facts, 95
Urey, Harold, 10, 12

van Fraassen, Bas, 109, 205
Vesalius, 116, 124, 158
Veblen, Thorstein, 143